LONDON MATHEMATICAL SOCIETY LECTURE NC

Managing Editor: Professor E. Süli, Mathematical Institute,
Woodstock Road, University of Oxford, Oxford OX2 6GG, United Kingdom

The titles below are available from booksellers, or from Cambridge University Press at
www.cambridge.org/mathematics

349 Model theory with applications to algebra and analysis I, Z. CHATZIDAKIS, D. MACPHERSON, A. PILLAY & A. WILKIE (eds)
350 Model theory with applications to algebra and analysis II, Z. CHATZIDAKIS, D. MACPHERSON, A. PILLAY & A. WILKIE (eds)
351 Finite von Neumann algebras and masas, A.M. SINCLAIR & R.R. SMITH
352 Number theory and polynomials, J. MCKEE & C. SMYTH (eds)
353 Trends in stochastic analysis, J. BLATH, P. MRTERS & M. SCHEUTZOW (eds)
354 Groups and analysis, K. TENT (ed)
355 Non-equilibrium statistical mechanics and turbulence, J. CARDY, G. FALKOVICH & K. GAWEDZKI
356 Elliptic curves and big Galois representations, D. DELBOURGO
357 Algebraic theory of differential equations, M.A.H. MACCALLUM & A.V. MIKHAILOV (eds)
358 Geometric and cohomological methods in group theory, M.R. BRIDSON, P.H. KROPHOLLER & I.J. LEARY (eds)
359 Moduli spaces and vector bundles, L. BRAMBILA-PAZ, S.B. BRADLOW, O. GARCÍA-PRADA & S. RAMANAN (eds)
360 Zariski geometries, B. ZILBER
361 Words: Notes on verbal width in groups, D. SEGAL
362 Differential tensor algebras and their module categories, R. BAUTISTA, L. SALMERÓN & R. ZUAZUA
363 Foundations of computational mathematics, Hong Kong 2008, F. CUCKER, A. PINKUS & M.J. TODD (eds)
364 Partial differential equations and fluid mechanics, J.C. ROBINSON & J.L. RODRIGO (eds)
365 Surveys in combinatorics 2009, S. HUCZYNSKA, J.D. MITCHELL & C.M. RONEY-DOUGAL (eds)
366 Highly oscillatory problems, B. ENGQUIST, A. FOKAS, E. HAIRER & A. ISERLES (eds)
367 Random matrices: High dimensional phenomena, G. BLOWER
368 Geometry of Riemann surfaces, F.P. GARDINER, G. GONZÁLEZ-DIEZ & C. KOUROUNIOTIS (eds)
369 Epidemics and rumours in complex networks, M. DRAIEF & L. MASSOULIÉ
370 Theory of p-adic distributions, S. ALBEVERIO, A.YU. KHRENNIKOV & V.M. SHELKOVICH
371 Conformal fractals, F. PRZYTYCKI & M. URBAŃSKI
372 Moonshine: The first quarter century and beyond, J. LEPOWSKY, J. MCKAY & M.P. TUITE (eds)
373 Smoothness, regularity and complete intersection, J. MAJADAS & A. G. RODICIO
374 Geometric analysis of hyperbolic differential equations: An introduction, S. ALINHAC
375 Triangulated categories, T. HOLM, P. JØRGENSEN & R. ROUQUIER (eds)
376 Permutation patterns, S. LINTON, N. RUŠKUC & V. VATTER (eds)
377 An introduction to Galois cohomology and its applications, G. BERHUY
378 Probability and mathematical genetics, N. H. BINGHAM & C. M. GOLDIE (eds)
379 Finite and algorithmic model theory, J. ESPARZA, C. MICHAUX & C. STEINHORN (eds)
380 Real and complex singularities, M. MANOEL, M.C. ROMERO FUSTER & C.T.C WALL (eds)
381 Symmetries and integrability of difference equations, D. LEVI, P. OLVER, Z. THOMOVA & P. WINTERNITZ (eds)
382 Forcing with random variables and proof complexity, J. KRAJÍČEK
383 Motivic integration and its interactions with model theory and non-Archimedean geometry I, R. CLUCKERS, J. NICAISE & J. SEBAG (eds)
384 Motivic integration and its interactions with model theory and non-Archimedean geometry II, R. CLUCKERS, J. NICAISE & J. SEBAG (eds)
385 Entropy of hidden Markov processes and connections to dynamical systems, B. MARCUS, K. PETERSEN & T. WEISSMAN (eds)
386 Independence-friendly logic, A.L. MANN, G. SANDU & M. SEVENSTER
387 Groups St Andrews 2009 in Bath I, C.M. CAMPBELL *et al* (eds)
388 Groups St Andrews 2009 in Bath II, C.M. CAMPBELL *et al* (eds)
389 Random fields on the sphere, D. MARINUCCI & G. PECCATI
390 Localization in periodic potentials, D.E. PELINOVSKY
391 Fusion systems in algebra and topology, M. ASCHBACHER, R. KESSAR & B. OLIVER
392 Surveys in combinatorics 2011, R. CHAPMAN (ed)
393 Non-abelian fundamental groups and Iwasawa theory, J. COATES *et al* (eds)
394 Variational problems in differential geometry, R. BIELAWSKI, K. HOUSTON & M. SPEIGHT (eds)
395 How groups grow, A. MANN
396 Arithmetic differential operators over the p-adic integers, C.C. RALPH & S.R. SIMANCA
397 Hyperbolic geometry and applications in quantum chaos and cosmology, J. BOLTE & F. STEINER (eds)
398 Mathematical models in contact mechanics, M. SOFONEA & A. MATEI
399 Circuit double cover of graphs, C.-Q. ZHANG
400 Dense sphere packings: a blueprint for formal proofs, T. HALES
401 A double Hall algebra approach to affine quantum Schur-Weyl theory, B. DENG, J. DU & Q. FU

402 Mathematical aspects of fluid mechanics, J.C. ROBINSON, J.L. RODRIGO & W. SADOWSKI (eds)
403 Foundations of computational mathematics, Budapest 2011, F. CUCKER, T. KRICK, A. PINKUS & A. SZANTO (eds)
404 Operator methods for boundary value problems, S. HASSI, H.S.V. DE SNOO & F.H. SZAFRANIEC (eds)
405 Torsors, tale homotopy and applications to rational points, A.N. SKOROBOGATOV (ed)
406 Appalachian set theory, J. CUMMINGS & E. SCHIMMERLING (eds)
407 The maximal subgroups of the low-dimensional finite classical groups, J.N. BRAY, D.F. HOLT & C.M. RONEY-DOUGAL
408 Complexity science: the Warwick master's course, R. BALL, V. KOLOKOLTSOV & R.S. MACKAY (eds)
409 Surveys in combinatorics 2013, S.R. BLACKBURN, S. GERKE & M. WILDON (eds)
410 Representation theory and harmonic analysis of wreath products of finite groups, T. CECCHERINI-SILBERSTEIN, F. SCARABOTTI & F. TOLLI
411 Moduli spaces, L. BRAMBILA-PAZ, O. GARCÍA-PRADA, P. NEWSTEAD & R.P. THOMAS (eds)
412 Automorphisms and equivalence relations in topological dynamics, D.B. ELLIS & R. ELLIS
413 Optimal transportation, Y. OLLIVIER, H. PAJOT & C. VILLANI (eds)
414 Automorphic forms and Galois representations I, F. DIAMOND, P.L. KASSAEI & M. KIM (eds)
415 Automorphic forms and Galois representations II, F. DIAMOND, P.L. KASSAEI & M. KIM (eds)
416 Reversibility in dynamics and group theory, A.G. O'FARRELL & I. SHORT
417 Recent advances in algebraic geometry, C.D. HACON, M. MUSTAŢĂ & M. POPA (eds)
418 The Bloch-Kato conjecture for the Riemann zeta function, J. COATES, A. RAGHURAM, A. SAIKIA & R. SUJATHA (eds)
419 The Cauchy problem for non-Lipschitz semi-linear parabolic partial differential equations, J.C. MEYER & D.J. NEEDHAM
420 Arithmetic and geometry, L. DIEULEFAIT *et al* (eds)
421 O-minimality and Diophantine geometry, G.O. JONES & A.J. WILKIE (eds)
422 Groups St Andrews 2013, C.M. CAMPBELL *et al* (eds)
423 Inequalities for graph eigenvalues, Z. STANIĆ
424 Surveys in combinatorics 2015, A. CZUMAJ *et al* (eds)
425 Geometry, topology and dynamics in negative curvature, C.S. ARAVINDA, F.T. FARRELL & J.-F. LAFONT (eds)
426 Lectures on the theory of water waves, T. BRIDGES, M. GROVES & D. NICHOLLS (eds)
427 Recent advances in Hodge theory, M. KERR & G. PEARLSTEIN (eds)
428 Geometry in a Frchet context, C.T.J. DODSON, G. GALANIS & E. VASSILIOU
429 Sheaves and functions modulo *p*, L. TAELMAN
430 Recent progress in the theory of the Euler and Navier-Stokes equations, J.C. ROBINSON, J.L. RODRIGO, W. SADOWSKI & A. VIDAL-LÓPEZ (eds)
431 Harmonic and subharmonic function theory on the real hyperbolic ball, M. STOLL
432 Topics in graph automorphisms and reconstruction (2nd Edition), J. LAURI & R. SCAPELLATO
433 Regular and irregular holonomic D-modules, M. KASHIWARA & P. SCHAPIRA
434 Analytic semigroups and semilinear initial boundary value problems (2nd Edition), K. TAIRA
435 Graded rings and graded Grothendieck groups, R. HAZRAT
436 Groups, graphs and random walks, T. CECCHERINI-SILBERSTEIN, M. SALVATORI & E. SAVA-HUSS (eds)
437 Dynamics and analytic number theory, D. BADZIAHIN, A. GORODNIK & N. PEYERIMHOFF (eds)
438 Random walks and heat kernels on graphs, M.T. BARLOW
439 Evolution equations, K. AMMARI & S. GERBI (eds)
440 Surveys in combinatorics 2017, A. CLAESSON *et al* (eds)
441 Polynomials and the mod 2 Steenrod algebra I, G. WALKER & R.M.W. WOOD
442 Polynomials and the mod 2 Steenrod algebra II, G. WALKER & R.M.W. WOOD
443 Asymptotic analysis in general relativity, T. DAUDÉ, D. HÄFNER & J.-P. NICOLAS (eds)
444 Geometric and cohomological group theory, P.H. KROPHOLLER, I.J. LEARY, C. MARTÍNEZ-PÉREZ & B.E.A. NUCINKIS (eds)
445 Introduction to hidden semi-Markov models, J. VAN DER HOEK & R.J. ELLIOTT
446 Advances in two-dimensional homotopy and combinatorial group theory, W. METZLER & S. ROSEBROCK (eds)
447 New directions in locally compact groups, P.-E. CAPRACE & N. MONOD (eds)
448 Synthetic differential topology, M.C. BUNGE, F. GAGO & A.M. SAN LUIS
449 Permutation groups and cartesian decompositions, C.E. PRAEGER & C. SCHNEIDER
450 Partial differential equations arising from physics and geometry, M. BEN AYED *et al* (eds)
451 Topological methods in group theory, N. BROADDUS, M. DAVIS, J.-F. LAFONT & I. ORTIZ (eds)
452 Partial differential equations in fluid mechanics, C.L. FEFFERMAN, J.C. ROBINSON & J.L. RODRIGO (eds)
453 Stochastic stability of differential equations in abstract spaces, K. LIU
454 Beyond hyperbolicity, M. HAGEN, R. WEBB & H. WILTON (eds)
455 Groups St Andrews 2017 in Birmingham, C.M. CAMPBELL *et al* (eds)

London Mathematical Society Lecture Note Series: 454

Beyond Hyperbolicity

Edited by

MARK HAGEN
University of Bristol

RICHARD WEBB
University of Cambridge

HENRY WILTON
University of Cambridge

CAMBRIDGE
UNIVERSITY PRESS

University Printing House, Cambridge CB2 8BS, United Kingdom

One Liberty Plaza, 20th Floor, New York, NY 10006, USA

477 Williamstown Road, Port Melbourne, VIC 3207, Australia

314–321, 3rd Floor, Plot 3, Splendor Forum, Jasola District Centre,
New Delhi – 110025, India

79 Anson Road, #06–04/06, Singapore 079906

Cambridge University Press is part of the University of Cambridge.

It furthers the University's mission by disseminating knowledge in the pursuit of education, learning, and research at the highest international levels of excellence.

www.cambridge.org
Information on this title: www.cambridge.org/9781108447294
DOI: 10.1017/9781108559065

First published 2019

Printed and bound in Great Britain by Clays Ltd, Elcograf S.p.A.

A catalogue record for this publication is available from the British Library

ISBN 978-1-108-44729-4 Paperback

Contents

Preface *page* ix

PART ONE LECTURES 1

1 Notes on coarse median spaces ***B.H. Bowditch*** 3

- 1.1 Introduction 3
- 1.2 Quasi-isometry invariants 4
- 1.3 Medians 8
- 1.4 Coarse median spaces 11
- 1.5 Surfaces 13
- 1.6 Asymptotic cones 18

2 Semihyperbolicity ***M.R. Bridson*** 25

- 2.1 The universe of finitely presented groups 27
- 2.2 Some key features of hyperbolic groups 30
- 2.3 Some properties of CAT(0) groups 37
- 2.4 Combings and semihyperbolicity 38
- 2.5 Languages and the complexity of normal forms 46
- 2.6 Examples 52
- 2.7 Algorithmic construction of classifying spaces 56
- 2.8 Cubulated groups and systolic groups 57
- 2.9 Subgroups 58
- 2.10 Containments 59

3 Acylindrically hyperbolic groups ***B. Barrett*** 65

- 3.1 Acylindrically hyperbolic groups 65
- 3.2 Small-cancellation theory 69
- 3.3 Dehn surgery 72
- 3.4 The extension problem 75
- 3.5 Acylindrically hyperbolic structures 78

PART TWO EXPOSITORY ARTICLES 81

4 A survey on Morse boundaries and stability *M. Cordes* 83
4.1 Generalizing hyperbolicity 83
4.2 Contracting and Morse boundaries 85
4.3 (Metric) Morse boundary and stability 95
4.4 Stable subgroups 106
4.5 A metrisable topology on the Morse boundary 110

5 What is a hierarchically hyperbolic space? *A. Sisto* 117
5.1 Heuristic discussion 120
5.2 Technical discussion 129

PART THREE RESEARCH ARTICLES 149

6 A counterexample to questions about boundaries, stability, and commensurability *J. Behrstock* 151
6.1 The construction 152
6.2 Properties 153
6.3 Applications 155
6.4 Further questions 157

7 A note on the acylindrical hyperbolicity of groups acting on CAT(0) cube complexes *I. Chatterji and A. Martin* 160
7.1 Introduction 160
7.2 Über-contractions and acylindrical hyperbolicity 164
7.3 Über-separated hyperplanes and the proof of Theorem 7.1.1 165
7.4 Proof of Theorem 7.1.5 170
7.5 Artin groups of type FC 173

8 Immutability is not uniformly decidable in hyperbolic groups *D. Groves and H. Wilton* 179

9 Sphere systems, standard form, and cores of products of trees *F. Iezzi* 186
9.1 Introduction 186
9.2 Spheres, partitions and intersections 189
9.3 Standard form for sphere systems, piece decomposition and dual square complexes 191
9.4 The core of two trees 197
9.5 The inverse construction 207
9.6 Consequences and applications 217

10 Uniform quasiconvexity of the disc graphs in the curve graphs ***K.M. Vokes*** 223
10.1 Introduction 223
10.2 Preliminaries 225
10.3 Exceptional cases 226
10.4 Proof of the main result 226

Preface

The *Beyond hyperbolicity* workshop was held at the University of Cambridge Centre for Mathematical Sciences from 20–24 June 2016, with the goal of examining various generalizations of Gromov-hyperbolicity that have recently assumed a prominent role in geometric group theory. The enormous success of the theory of Gromov-hyperbolic spaces and groups in the 30 years since their introduction has inspired geometric group theorists to go "beyond hyperbolicity", i.e., to study groups which are not hyperbolic by exploiting the vestiges of hyperbolicity that they nonetheless exhibit. This paradigm is exemplified by theories that directly generalise hyperbolicity (e.g. relative, acylindrical, and hierarchical hyperbolicity, or the theory of coarse median spaces), and also by ideas reminiscent of the thin-triangle condition that defines a hyperbolic space (e.g. median spaces and various flavours of nonpositive curvature). A major goal of the workshop — and one of our primary aims in creating this volume — was to survey this rich ecosystem of ideas and put them into conversation with one another.

Since a good part of the utility of the theory of hyperbolic groups comes from the notion of the Gromov boundary of a hyperbolic space, it was also our goal to describe the generalisations of the Gromov boundary arising in recent work, notably the contracting boundary, Morse boundary, and hierarchically hyperbolic boundary. Another goal was to look closely at important examples of groups exhibiting various "hyperbolic-like" features: right-angled Artin and other cubulated groups, mapping class groups of surfaces, outer automorphism groups of free groups, et cetera.

The workshop was organised around three mini-courses, whose lecture notes form the basis of the first three articles in this volume: Brian Bowditch (U. Warwick) lectured on coarse median spaces, Martin Bridson (U. Oxford) on semihyperbolicity, and Denis Osin (Vanderbilt U.) on acylindrical hyperbolicity. In addition to the mini-courses, there were fifteen lectures encompassing a broad range of recent developments in geometric group theory.

The present volume is based on material addressed in the workshop and aims to provide both a snapshot of the present state of this important branch of geometric group theory and also a reference for those wishing to acquaint themselves with the salient parts of the field. Therefore, in addition to the expository articles based on the mini-courses, we have also included expository articles on two extra topics of current interest: Morse boundaries, and hierarchical hyperbolicity. In addition to the expository articles, there are several research articles representing recent contributions to the theory of groups exhibiting hyperbolic features.

We are very grateful to all of the authors for producing such excellent articles, the anonymous referees for their essential comments, and Benjamin Barrett for meticulously transcribing and beautifully illustrating the mini-course lectures, on which the first three articles in this volume are based. We are also very grateful to those who made the workshop such an exciting and stimulating event: the speakers, the more than 70 participants, the Centre for Mathematical Sciences, and Selwyn College. We finally acknowledge the financial assistance of the Engineering and Physical Sciences Research Council (GRN EP/1003843/2 and EP/L026481/1), which made the workshop possible.

Mark Hagen
Richard Webb
Henry Wilton

PART ONE

LECTURES

1

Notes on coarse median spaces

Brian H. Bowditch
Mathematics Institute
University of Warwick
Coventry CV4 7AL
United Kingdom

Abstract

These are notes from a mini-course lectured by Brian H. Bowditch on coarse median spaces given at *Beyond Hyperbolicity* in Cambridge in June 2016.

1.1 Introduction

These lecture notes give a brief summary of the notion of a "coarse median space" as defined in [Bo1] and motivated by the centroid construction given in [BM2]. The basic idea is to capture certain aspects of the large-scale "cubical" structure of various naturally-occurring spaces. Thus, a coarse median space is a geodesic metric space equipped with a ternary "coarse median" operation, defined up to bounded distance, and satisfying a couple of simple axioms. Roughly speaking, these require that any finite subset of the space can be embedded in a finite CAT(0) cube complex in such a way that the coarse median operation agrees, up to bounded distance, with the natural combinatorial median in such a complex. One could express everything in terms of CAT(0) cube complexes, but it is more convenient to formulate it in terms of median algebras (which are essentially equivalent structures for finite sets). One can apply this notion to finitely generated groups via their Cayley graphs. Examples of coarse median spaces include Gromov hyperbolic spaces, mapping class groups and Teichmüller spaces of compact surfaces, right-angled Artin groups and geometrically finite kleinian groups in any dimension. The notion is useful for establishing certain results such as coarse rank and quasi-isometric rigidity for such spaces.

In Sections 1.2 and 1.3 we review some of the background to coarse geometry and to median algebras respectively. In Section 1.4 we combine these ideas to introduce the notion of a coarse median space. In Section 1.5 we discuss the geometry of the mapping class groups and Teichmüller spaces. In Section 1.6 we outline how the coarse median property is applied to such spaces via asymptotic cones.

I thank Benjamin Barrett for his excellent work in preparing these notes, based on a mini-course I gave at the meeting "Beyond Hyperbolicity" in Cambridge in June 2016. I take responsibility for any errors introduced by my subsequent editing and elaborations. I thank the organisers, Mark Hagen, Richard Webb and Henry Wilton, for their invitation to speak at the meeting.

1.2 Quasi-isometry invariants

We begin by making some basic definitions which describe the types of spaces we wish to discuss.

Let (X, ρ) be a metric space.

Definition 1.2.1 A *geodesic* in X is a path whose length is equal to the distance between its endpoints. We say that X is a *geodesic metric space* if every pair of points in X is the pair of endpoints of some geodesic.

All of the metric spaces of interest in this paper will be geodesic spaces (though we only make this hypothesis where we need it).

Definition 1.2.2 A geodesic space X is *proper* if it is complete and locally compact.

(This is equivalent to saying that all closed bounded subsets of X are compact.)

Definition 1.2.3 Let (X, ρ) and (X', ρ') be geodesic metric spaces. We say that a map $\phi\colon X \to X'$ is *coarsely-Lipschitz* if there exist constants $k_1, k_2 \geq 0$ such that

$$\rho'(\phi(x), \phi(y)) \leq k_1 \rho(x, y) + k_2$$

for any x and y in X.

We say that ϕ is a *quasi-isometric embedding* if it is coarsely-Lipschitz and there also exist constants $k_1', k_2' \geq 0$ such that

$$\rho(x, y) \leq k_1' \rho'(\phi(x), \phi(y)) + k_2'$$

for any x and y in X.

We say that ϕ is a *quasi-isometry* if it is a quasi-isometric embedding and there also exists a constant $k_3 \geq 0$ such that $X' = N(\phi(X), k_3)$; that is, X' is equal to the k_3-neighbourhood of the image of ϕ. In other words the image of ϕ is cobounded.

Note that in this definition we do not assume that the map ϕ is continuous.

Given geodesic spaces, X, Y, we write $X \leq Y$ if there exists a quasi-isometric embedding $X \to Y$, and $X \sim Y$ if there exists a quasi-isometry $X \to Y$. Then the relations $\leq$ and $\sim$ are both reflexive and transitive and $\sim$ is also symmetric. However $\leq$ is not antisymmetric: there exist spaces X and Y such that $X \leq Y$ and $Y \leq X$ but $X \not\sim Y$. For example, consider the following subsets of the euclidean plane, $\mathbb{R}^2$, given by

$$\begin{aligned}\{(x,y) \mid x, y \geq 0\} &\hookrightarrow \{(x,y) \mid (x \geq 0 \text{ and } y \geq 0) \text{ or } x = 0\} \\ &\hookrightarrow \{(x,y) \mid x \geq 0\} \\ &\sim \{(x,y) \mid x, y \geq 0\}\end{aligned}$$

in the induced path metrics. It is not hard to show that the intermediate spaces are not quasi-isometric to each other.

Example 1.2.4 For any $n \geq 1$, we have $[0, \infty)^n \sim \mathbb{R}^{n-1} \times [0, \infty)$. (Indeed, one can see easily that these spaces are bi-Lipschitz equivalent.) This half-space will appear again; we denote it H^n. Note that it is equipped with the restriction of the euclidean metric (not the hyperbolic metric).

Definition 1.2.5 Let a group Γ act on a proper geodesic metric space X by isometries. The action is *properly discontinuous* if for any compact subset K of X the set

$$\{g \in \Gamma \mid gK \cap K \neq \emptyset\}$$

is finite. (In this case the quotient space X/Γ is Hausdorff.)

The action is *cocompact* if X/Γ is compact.

When the action is cocompact, one can show that Γ must be finitely generated.

The geometry of a group is related to the geometry of the spaces on which it acts by the following theorem.

Theorem 1.2.6 (Švarc-Milnor) *If Γ acts on proper geodesic metric spaces X and X' properly discontinuously, cocompactly and by isometries, then $X \sim X'$. (Indeed, we can take the quasi-isometry to be equivariant.)*

Example 1.2.7 The action of a group Γ by left-translation on its Cayley graph $\Delta(\Gamma)$ with respect to any finite generating set is properly discontinuous and cocompact. It follows by Theorem 1.2.6 that any two such Cayley graphs for the same group are quasi-isometric.

Note: throughout this paper, unless otherwise stated, we assume that any connected graph is equipped with the combinatorial path metric, which assigns unit length to each edge.

Remark 1.2.8 We can often assume quasi-isometries to be continuous. For example, if $I \subset \mathbb{R}$ is an interval, then any quasi-isometric embedding $\phi: I \to X$ is within a bounded distance of a continuous map, and such a map is automatically also a quasi-isometric embedding. We refer to such a map as a *quasi-geodesic.*

Theorem 1.2.9 $\mathbb{R}^2 \not\leq \mathbb{R}$.

Proof Suppose for contradiction that $\phi: \mathbb{R}^2 \to \mathbb{R}$ is a quasi-isometric embedding. Without loss of generality, ϕ is continuous (since a simple argument shows that it can always be approximated up to bounded distance by a continuous map). Let $S \subset \mathbb{R}^2$ be a round circle of large radius centred at the origin. By the Intermediate Value Theorem there exists x in S such that $\phi(x) = \phi(-x)$, which gives a contradiction, provided we choose the radius sufficiently large in relation to the quasi-isometric parameters. □

In fact the same argument (choosing the centre of the circle appropriately) shows that $H^2 \not\leq \mathbb{R}$. Moreover, replacing the Intermediate Value Theorem with the Borsuk–Ulam theorem, one can see that $\mathbb{R}^{n+1} \not\leq \mathbb{R}^n$ for any n, and therefore $\mathbb{R}^m \sim \mathbb{R}^n$ only when $m = n$. Indeed one can see that $H^{n+1} \not\leq \mathbb{R}^n$. By related arguments one can also show that any quasi-isometric embedding of $\mathbb{R}^n$ into itself in necessarily a quasi-isometry.

Definition 1.2.10 If X is a geodesic space, the *euclidean rank* of X E-rk$(X) \in \mathbb{N} \cup \{\infty\}$ is defined to be the maximum n such that $\mathbb{R}^n \leq X$. The *half-space rank* of X, H-rk(X), is defined to be the maximum n such that $H^n \leq X$.

Clearly, H-rk$(X) - 1 \leq$ E-rk$(X) \leq$ H-rk(X). These ranks are quasi-isometry invariants.

Note that, by the above observations, we have E-rk$(\mathbb{R}^n)$ = H-rk$(\mathbb{R}^n) = n$ and E-rk$(H^n) + 1$ = H-rk$(H^n) = n$.

Definition 1.2.11 A map $f\colon [0,\infty) \to [0,\infty)$ is an *isoperimetric bound* for X if there exists a constant k such that if $\gamma\colon S^1 \to X$ is any curve, we can cut γ into at most $f(\text{length}(\gamma))$ loops of length at most k.

(More formally, we can extend f to a map of the 1-skeleton of a cellulation of the disc, with boundary S^1, such that the length of the f-image of the boundary of any 2-cell has length at most k.)

The rate of growth of the isoperimetric bound is a quasi-isometry invariant. (Here the "growth rate" is interpreted up to linear bounds: we allow for linear reparametrisation of the domain and range of f.) In particular, we can talk about spaces with linear, quadratic and exponential isoperimetric bounds, et cetera.

A central notion in the subject is that of *Gromov hyperbolicity* [G1]. There are numerous equivalent definitions, among which we choose the following.

Definition 1.2.12 A geodesic metric space X is *hyperbolic* if there exists a constant k such that for any geodesic triangle in X, there exists a point m in X within distance k of each of the three sides of the triangle. (A "geodesic triangle" consists of three geodesic segments — its "sides" — cyclically connecting three points.)

It turns out that, up to bounded distance, m depends only on the vertices of the triangle, so if x, y and z are the vertices then we write $m = m(x,y,z)$.

This definition is quasi-isometry invariant. Moreover, Gromov showed that X is hyperbolic if and only if it has a linear isoperimetric bound. We note also the following geometric properties of hyperbolic spaces.

1 Hyperbolic metric spaces satisfy a *Morse Lemma*: any quasi-geodesic is close to any geodesic joining its end points. More precisely, the Hausdorff distance between them depends only on the quasi-isometry constants and the hyperbolicity constant k.
2 Hyperbolic metric spaces can be well approximated by trees: there exists a function $h\colon \mathbb{N} \to \mathbb{N}$ such that if X is k-hyperbolic and $A \subset X$ is a finite subset of cardinality at most p, there exists a tree $\tau \subset X$ with $A \subset \tau$ such that for any x and y in A, $\rho_\tau(x,y) \le \rho(x,y) + kh(p)$. Here ρ_τ denotes the induced path-metric on τ. (In this case we are allowing the edges of τ to have differing lengths.) Note that, using the Morse Lemma, it follows that the arc in τ from x to y is a bounded Hausdorff distance from any geodesic in X from x to y.

Also note that if X is hyperbolic, then $\text{H-rk}(X) \leq 1$.

Definition 1.2.13 Let a group Γ act on a geodesic space X by isometries. We say that the action is *quasi-isometrically rigid* if for any quasi-isometry $\phi: X \to X$ there exists $g \in \Gamma$ such that $\rho(gx, \phi(x)) \leq C$ for some constant C depending only on the quasi-isometry constants of the map.

When the group Γ is understood, we will express this by saying that X is "quasi-isometrically rigid".

1.3 Medians

We describe the basic properties of a median algebra and how they relate to CAT(0) cube complexes. Some basic references for median algebras are [BaH, Ro, Ve]. Some further discussion, relevant to these notes, is given in [Bo1, Bo4]. CAT(0) complexes are discussed, for example, in [BrH]. We can view a CAT(0) complex combinatorially as a simply-connected complex built out of cubes such that the link of every vertex is a flag simplicial complex. They are usually equipped with a euclidean (CAT(0)) cubical structure, though it is more natural to consider the ℓ^1 metric in the present context.

Let M be a set and let $\mu: M^3 \to M$ be a ternary operation. (Intuitively, we think of μ as mapping points a, b and c in M to a point "between a, b and c".)

The standard definition of a median algebra is simple, but somewhat formal and perhaps unintuitive.

Definition 1.3.1 (M, μ) is a *median algebra* if for any a, b, c, d and e in M,

(M1) $\mu(a,b,c) = \mu(b,a,c) = \mu(b,c,a)$,
(M2) $\mu(a,a,b) = a$ and
(M3) $\mu(a,b,\mu(c,d,e)) = \mu(\mu(a,b,c),\mu(a,b,d),e)$.

Given a and b in M we write $[a,b]_\mu = \{x \in M \mid \mu(a,b,x) = x\}$, which we abbreviate to $[a,b]$ if the choice of function μ is clear from context. The set $[a,b]$ is called the *interval* between a and b.

The notion of a median algebra can equivalently, and perhaps more intuitively, be formulated in terms of intervals. This follows from work of Sholander [Sho]. (See [Bo4] for some elaboration.)

Lemma 1.3.2 *Let M be a median algebra. The interval operation $[\cdot,\cdot]$ satisfies the following properties for any a, b, c in M:*

(I1) $[a,a] = \{a\}$,
(I2) $[a,b] = [b,a]$,
(I3) $c \in [a,b] \implies [a,c] \subset [a,b]$, *and*
(I4) there exists d (depending on a, b and c) such that $[a,b] \cap [b,c] \cap [c,a] = \{d\}$.

In property (I4) we can set $d = \mu(a,b,c)$.

We can alternatively view properties (I1)–(I4) as axioms, and we have the following converse for any set M.

Theorem 1.3.3 *[Sho] Given a map $[\cdot,\cdot]$ from M^2 to the power set $\mathcal{P}(M)$ satisfying axioms (I1)–(I4) above, there exists a map $\mu\colon M^3 \to M$ such that (M,μ) is a median algebra and $[\cdot,\cdot] = [\cdot,\cdot]_\mu$. In fact, we can set $\mu(a,b,c)$ to be the element d given in axiom (I4).*

Example 1.3.4 We give some examples of median algebras.

1 Let M be the two-point set $\{0,1\}$. Then there is a unique median algebra structure on M given by $\mu(0,0,0) = 0$, $\mu(0,0,1) = 0$, $\mu(0,1,1) = 1$, $\mu(1,1,1) = 1$ etc. (In other words μ represents the "majority vote".)
2 If M_1 and M_2 are median algebras then so is $M_1 \times M_2$, with the median defined separately on each co-ordinate.
3 Combining the previous two examples, the "n-cube" $\{0,1\}^n$ has a natural median algebra structure. One can show that any finite median algebra is a subalgebra of such a cube.
4 Trees are median algebras. Define the median of three points to be the centre of the tripod spanned by those points. Here a "tree" can be interpreted as a simplicial tree, or more generally any $\mathbb{R}$-tree. This includes the case of $\mathbb{R}$ itself: here the median of three points is just the point that lies between the other two.
5 Given any set X define a median on its power set $\mathcal{P}(X)$ by:

$$\begin{aligned}\mu(A,B,C) &= (A \cup B) \cap (B \cup C) \cap (C \cup A)\\ &= (A \cap B) \cup (B \cap C) \cup (C \cap A)\end{aligned}$$

for $A,B,C \subset X$. Then $(\mathcal{P}(X),\mu)$ is a median algebra.
6 The previous example generalises to any distributive lattice, with the median defined by a similar formula, using meets and joins in place of intersections and unions.

7 Let Δ be a CAT(0) cube complex. Its vertex set $V(\Delta)$ can be made into a median algebra as follows. Let ρ be the combinatorial path metric on the 1-skeleton of Δ. Then given $a, b \in V(\Delta)$ let $[a,b]_\rho = \{x \in M : \rho(a,b) = \rho(a,x) + \rho(x,b)\}$. This definition satisfies axioms (I1)–(I4) above, so by Theorem 1.3.3 there exists a median algebra structure $\mu : V(\Delta)^3 \to V(\Delta)$ such that $[a,b]_\mu = [a,b]_\rho$.

8 $\mathbb{R}^n$ with the ℓ^1 metric, ρ. Here one defines the median similarly as in the previous example. This is median-isomorphic to the direct product of n copies of $\mathbb{R}$.

9 Similarly, CAT(0) cube complexes with the ℓ^1 metric (that is the path-metric obtained by putting the ℓ^1 metric on each cube). In this case, the vertex set is a subalgebra (that is, closed under μ).

10 More generally, a median metric space: that is any metric space (X, ρ) such that $[a,b]_\rho \cap [b,c]_\rho \cap [c,a]_\rho$ is a singleton for all $a, b, c \in X$ (which gives us the median of a, b, c). Note that this is just axiom (I4) in Theorem 1.3.3. Axioms (I1)–(I3) follow immediately from the metric space axioms.

A subset B of a median algebra M is a *subalgebra* if it is closed under μ. We write $B \le M$. For any $A \subset M$, $\langle A \rangle \le M$ is the subalgebra generated by A; that is, the intersection of all subalgebras of M containing A.

We say that a subset $C \subset M$ is *convex* if $[a,b] \subset C$ whenever $a, b \in C$. We note that convex sets are subalgebras, and that intervals themselves are convex.

The following are two basic facts about median algebras.

Theorem 1.3.5

1 Let M be a median algebra, and let $A \subset M$ with $|A| \le p < \infty$. Then $|\langle A \rangle| \le 2^{2^p}$.

2 Any finite median algebra is canonically the vertex set of a CAT(0) *cube complex.*

Note that these give rise to a third equivalent way of defining a median algebra: it is a set equipped with a ternary operation such that any finite subset is contained in another finite subset, closed under this operation, and isomorphic to the median structure on a finite CAT(0) cube complex.

In particular, in dealing with any finite subset of a median algebra, we can often just pretend we are living in a CAT(0) cube complex.

Definition 1.3.6 Define the *median rank* of M, M-rk(M), to be the maximum n such that $\{0,1\}^n \le M$, so M-rk$(M) \in \mathbb{N} \cup \{\infty\}$.

For example, M-rk($\mathbb{R}^n$) = n, and if Δ is a CAT(0) cube complex then one can check that M-rk(Δ) = M-rk($V(\Delta)$) = dim(Δ) (that is, the standard notion of dimension — the maximal dimension of a cubical cell).

We now state two theorems about median metric spaces which will be useful for the discussion in Section 1.6.

Theorem 1.3.7 *[Bo1] Let M be a connected, locally convex topological median algebra of rank at most $n < \infty$. Then the locally compact dimension of M (i.e. the maximum topological dimension of a locally compact subset of M) is equal to the median rank* M-rk(M).

(For locally compact spaces, all of the standard definitions of topological dimension are equivalent [E]. For definiteness, we could take to mean covering dimension.)

Here a "topological median algebra" is simply one equipped with a topology with respect to which the median operation is continuous. It is "locally convex" if every point has a base of convex neighbourhoods. This is satisfied in the cases of interest here. For example, the hypotheses of the theorem hold in any finite-rank connected median metric space (as defined in Example 1.3.4 (10)).

Theorem 1.3.8 ([Bo4]) *If (M, ρ) is a connected complete finite-rank median metric space then there exists a canonical bi-Lipschitz-equivalent metric σ_ρ on M such that (M, σ_ρ) is* CAT(0).

We remark that, under the same hypotheses, one can also put a canonical bi-Lipschitz equivalent injective metric on M [Mi, Bo8].

In particular, it follows (from either the CAT(0) or injective metric) that M is contractible.

1.4 Coarse median spaces

Let (Λ, ρ) be a geodesic space.

Definition 1.4.1 A map $\mu\colon \Lambda^3 \to \Lambda$ is a *coarse median* if

(C1) There exist k and l such that for any $a, b, c, a', b', c' \in M$,

$$\rho(\mu(a,b,c), \mu(a',b',c')) \le k(\rho(a,a') + \rho(b,b') + \rho(c,c')) + l.$$

(C2) There exists $h\colon \mathbb{N} \to [0,\infty)$ such that if A is a subset of Λ containing at most p points, then there exists a finite median algebra (Π, μ_Π) and maps

$$A \xrightarrow{\pi} \Pi \xrightarrow{\lambda} \Lambda$$

such that $\rho(a, \lambda\pi a) \le h(p)$ for all $a \in A$, and

$$\rho\left(\mu(\lambda a, \lambda b, \lambda c), \lambda\mu_\Pi(a, b, c)\right) \le h(p)$$

for all a, b and c in Π.

We say that Λ has *coarse (median) rank* at most n if (given some fixed map h) we can always take $\text{M-rk}(\Pi) \le n$. The coarse rank (i.e. the minimal such n) is denoted $\text{C-rk}(\Lambda)$.

Less formally, (C1) says that the coarse median is coarsely-Lipschitz, and (C2) says that, on finite sets, it looks like the median on a finite CAT(0) cube complex up to bounded distance. We remark that an equivalent set of axioms for a coarse median space has recently been described in [NWZ].

The existence of a coarse median on a geodesic space is a quasi-isometry invariant. We say that a finitely generated group is *coarse median* if some (hence any) Cayley graph with respect to a finite generating set admits a coarse median.

We give some examples.

Example 1.4.2

1 CAT(0) cube complexes (so right-angled Artin groups are coarse median groups).
2 Hyperbolic spaces: these spaces have coarse rank at most 1. (This follows from approximation of the space by a tree). In fact any coarse median space with coarse rank at most 1 is hyperbolic [Bo1] (see [NWZ] for a more direct proof).
3 The property is closed under taking direct products and relative hyperbolicity [Bo2].
4 From this it follows that any geometrically finite kleinian group (in any dimension) is coarse median. So are limit groups, as defined by Sela.
5 Mapping class groups, Teichmüller space in either the Teichmüller metric or the Weil–Petersson metric [Bo6, Bo5, Bo7], and the separating curve graphs [Vo]. (See Section 1.5).
6 Any hierarchically hyperbolic space [BHS1, BHS2].

The following is fairly easy to see, and only requires axiom (C1) [Bo1].

Theorem 1.4.3 *Any coarse median space satisfies a quadratic isoperimetric bound.*

One can also show the following [Bo1].

Theorem 1.4.4 *If* Λ *is a coarse median space, then* $\text{H-rk}(\Lambda) \leq \text{M-rk}(\Lambda)$.

We will outline how this is proven in Section 1.6. We will first elaborate on its consequences for the mapping class groups and Teichmüller space in the next section.

1.5 Surfaces

Let Σ be a compact orientable surface. Let $g(\Sigma)$ be its genus, $p(\Sigma)$ be the number of boundary components of Σ, and define the *complexity* of Σ to be

$$\xi(\Sigma) = 3g(\Sigma) - 3 + p(\Sigma).$$

This is the maximum number of disjoint curves (i.e. essential non-peripheral simple closed curves up to homotopy) that one can embed in Σ. We usually assume that $\xi(\Sigma) \geq 2$. We will denote the topological type of a surface of genus g and p boundary components by $S_{g,p}$.

Recall that the mapping class group, $\text{Map}(\Sigma)$, can be defined as the group of self-homeomorphisms of Σ defined up to homotopy (or, equivalently, isotopy). This is a finitely presented group. For future reference, we note that $\mathbb{Z}^{\xi} \leq \text{Map}(\Sigma)$. For example take the subgroup generated by Dehn twists around any maximal collection of disjoint simple closed curves (that is a "pants decomposition" of Σ).

We will focus on four particular spaces on which the mapping class group acts, namely, the marking graph $\mathcal{M}$, the curve graph $\mathcal{C}$, and Teichmüller space with the Teichmüller or Weil–Petersson metric, respectively denoted $\mathcal{T}$ and $\mathcal{W}$.

These spaces are interrelated. In fact, there are coarsely-Lipschitz $\text{Map}(\Sigma)$-equivariant maps

$$\mathcal{M} \longrightarrow \mathcal{T} \longrightarrow \mathcal{W} \longrightarrow \mathcal{C}.$$

natural up to bounded distance.

We proceed to describe these spaces in more detail. We begin with the marking graph. We write $\iota(\alpha, \beta)$ for the intersection number of two curves α, β (that is the minimal cardinality of $|\alpha \cap \beta|$ among realisations in Σ).

Definition 1.5.1 A set, a, of curves in Σ is said to *fill* Σ if for any curve γ in Σ, there is some $\alpha \in a$ with $\iota(\gamma, \alpha) \neq 0$. (Less formally, this says that a cuts Σ into discs and peripheral annuli.) A *marking* is a set, a, of curves in Σ that fills Σ and such that for all $\alpha, \beta \in a$, $\iota(\alpha, \beta) \leq 100$. The *marking graph*, $\mathcal{M}$, has vertex set $V(\mathcal{M})$ equal to the set of markings of Σ, and where two markings a and b in $\mathcal{M}$ are deemed adjacent if for any $\alpha \in a$ and $\beta \in b$ we have $\iota(\alpha, \beta) \leq 10000$.

(Here "100" and "10000" could be interpreted to mean any two sufficiently large numbers.)

The graph $\mathcal{M}$ is connected and $\mathrm{Map}(\Sigma)$ acts on $\mathcal{M}$ properly discontinuously and cocompactly. In particular, by Theorem 1.2.6 we see that $\mathcal{M}$ is quasi-isometric to (any Cayley graph of) $\mathrm{Map}(\Sigma)$. (A different definition is given in [MM2]. The notion is quite robust — any two sensible definitions will give equivariantly quasi-isometric graphs. It will not matter to us here which variation is chosen.)

One can show that the subgroup $\mathbb{Z}^{\xi}$ of $\mathrm{Map}(\Sigma)$ generated by Dehn twists is quasi-isometrically embedded. (This means that any orbit of this group in $\mathcal{M}$ is quasi-isometrically embedded.) From this we see that $\mathbb{R}^{\xi} \preceq \mathcal{M}$. In other words, $\text{E-rk}(\mathcal{M}) \geq \xi$.

By a *Dehn twist flat* in $\mathcal{M}$ we will mean an $\mathbb{Z}^{\xi}$-orbit of this type, where the orbits are chosen to be uniformly quasi-isometrically embedded. Uniformity is possible since there are only finitely many conjugacy classes of subgroups of this type in $\mathrm{Map}(\Sigma)$. (Not all quasi-isometric embeddings of $\mathbb{R}^{\xi}$ into $\mathcal{M}$ arise from Dehn twist flats, however.)

Definition 1.5.2 We define the *curve graph*, $\mathcal{C}$, of Σ. Its vertex set $V(\mathcal{C})$ is the set of curves on Σ; two curves are deemed to be adjacent if they can be homotoped to be disjoint.

The curve graph is connected whenever $\xi(\Sigma) \geq 2$. In fact, the following result is central to the whole subject.

Theorem 1.5.3 *(Masur–Minsky) [MM1] $\mathcal{C}$ is hyperbolic.*

Recall that the *Teichmüller space* of Σ is the space of marked finite-area hyperbolic structures on $\Sigma - \partial\Sigma$, defined up to isotopy. (See, for example, [IT].) As a topological space it is homeomorphic to $\mathbb{R}^{2\xi}$, though we are interested here in its (large-scale) geometry. It admits many interesting metrics; for example, the Teichmüller metric and the Weil–Petersson metric as mentioned above.

The spaces $\mathcal{T}$ and $\mathcal{W}$ have a somewhat different structure. Notably, $\mathcal{T}$ is complete, whereas $\mathcal{W}$ is not. The basic reason behind this can be thought of as follows. Take any essential non-peripheral simple closed curve on the surface. One can form a path in Teichmüller space by shrinking the length of the curve, while keeping the hyperbolic structure on the remainder of the surface approximately constant. In this way the surface develops an annular "Margulis tube", with our curve as its core. As the length of this curve tends to 0, the length of the tubes tends to ∞. We thus get a properly embedded path in Teichmüller space. In $\mathcal{T}$, this process takes an infinite amount of effort, and the path has infinite length. However, in $\mathcal{W}$ only a finite amount of effort is needed to pull the surface apart in this way, and the path has finite length. (See [W].)

In fact, one can take a maximal collection of disjoint curves in Σ and shrink these independently of each other. Since there are ξ such curves, this gives a proper map of $[0,\infty)^\xi$ into Teichmüller space. One can show that the map $[0,\infty) \to \mathcal{T}$ is a quasi-isometric embedding. Since $[0,\infty)^\xi \sim H^\xi$, we see that $H^\xi \preceq \mathcal{T}$. In other words, $\text{H-rk}(\mathcal{T}) \geq \xi$.

In $\mathcal{W}$ however, the image of this map is bounded, so we don't achieve very much by this. Instead, write

$$\xi_0 = \xi_0(\Sigma) = \lfloor (\xi(\Sigma) + 1)/2 \rfloor.$$

The significance of this number is that we can cut Σ into ξ_0 pieces, each of complexity at least 1 (that means each contains an $S_{0,4}$ or an $S_{1,1}$). One can now deform the hyperbolic structures on these pieces independently, and by taking an appropriate bi-infinite path of such deformations in each component, we get a proper map of $\mathbb{R}^{\xi_0}$ into $\mathcal{W}$. One can also show that such a map is a quasi-isometric embedding. Therefore, $\text{E-rk}(\mathcal{W}) \geq \xi_0$.

In fact, one has equality when $\xi(\Sigma) \geq 2$, as the following theorem clarifies.

Theorem 1.5.4

1 (Behrstock, Minsky, Hamenstädt) [BM1, H] $\text{H-rk}(\mathcal{M}) = \text{E-rk}(\mathcal{M}) = \xi$.
2 (Eskin, Masur, Rafi) [EMR1] $\text{H-rk}(\mathcal{T}) = \xi$.
3 (Eskin, Masur, Rafi) [EMR1] $\text{H-rk}(\mathcal{W}) = \text{E-rk}(\mathcal{W}) = \xi_0$.

The remaining issue regarding $\text{E-rk}(\mathcal{T})$ is resolved by the following.

Theorem 1.5.5 *[Bo5]* $\mathbb{R}^\xi \preceq \mathcal{T}$ *if and only if* $g \leq 1$ *or* $\Sigma = S_{2,0}$.

We will outline later how Theorem 1.5.4 can also be derived from the coarse median property.

Recall that Map(Σ) acts naturally on $\mathcal{M}$, $\mathcal{T}$, $\mathcal{W}$ and $\mathcal{C}$. We briefly review some rigidity results for these actions.

Theorem 1.5.6

1 *(Behrstock, Kleiner, Minsky, Mosher, Hamenstädt) [BKMM, H] $\mathcal{M}$ is quasi-isometrically rigid.*
2 *(Eskin, Masur, Rafi, Bowditch) [EMR2, Bo5] $\mathcal{T}$ is quasi-isometrically rigid.*
3 *(Bowditch) [Bo7] $\mathcal{W}$ is quasi-isometrically rigid if $g(\Sigma) + p(\Sigma) \geq 7$.*
4 *(Rafi, Schleimer) [RS] $\mathcal{C}$ is quasi-isometrically rigid.*

It is a relatively simple matter to account for the low complexity cases ($\xi \leq 1$), so this give a compete answer for $\mathcal{M}$, $\mathcal{T}$ and $\mathcal{C}$. However, [Bo7] leaves unresolved about a dozen cases for $\mathcal{W}$.

We also have the following analogue of the cohopfian property (*cf.* the case of $\mathbb{R}^n$ discussed in Section 1.2).

Theorem 1.5.7 *[Bo6] Any quasi-isometric embedding of $\mathcal{M}$ into itself is a quasi-isometry.*

In fact, this is achieved by giving another proof of quasi-isometric rigidity of $\mathcal{M}$, but only using the weaker hypothesis that our map is a quasi-isometric embedding. (I do not know whether a similar statement holds for any of the other cases: $\mathcal{T}$, $\mathcal{W}$ or $\mathcal{C}$.)

It is time to explain how the coarse median property is brought into play.

Theorem 1.5.8 *[Bo6, Bo5, Bo7]. $\mathcal{M}$ and $\mathcal{T}$ are coarse median of rank ξ, $\mathcal{W}$ is coarse median of rank ξ_0 and $\mathcal{C}$ is coarse median of rank 1.*

(Of course, the last statement about $\mathcal{C}$ does not tell us anything essentially new — it follows directly from Theorem 1.5.3.)

From Theorem 1.4.3 it follows that each of these spaces satisfies a quadratic isoperimetric bound. So, for example, we recover the fact, due to Mosher [Mo], that the mapping class group has quadratic Dehn function. For $\mathcal{W}$ and $\mathcal{C}$ this follows respectively from the CAT(0) property and hyperbolicity of these spaces. The fact this holds for the Teichmüller metric appears to be new, though an independent proof has been announced by Kapovich and Rafi [KR].

Note that one can now also recover Theorem 1.5.4 using Theorem 1.4.4. (Another proof for $\mathcal{T}$ is given in [Du2], and for $\mathcal{T}$ and $\mathcal{W}$ in [BHS1].)

Theorem 1.5.5 still requires some more work, which we will not describe here.

As for the rigidity results, there is still quite a bit more to be done, but we will briefly discuss some of the ingredients in Section 1.6.

We spend the remainder of this section briefly describing how the coarse median structure arises in these situations. First, we consider the case of $\mathcal{M}$. This is based on the centroid construction of [BM2], which uses the notion of subsurface projection of Masur and Minsky [MM2].

Let $\mathcal{S}$ be the set of π_1-injective subsurfaces of Σ up to homotopy. For technical reasons it is helpful to exclude surfaces homeomorphic to $S_{0,1}$ or $S_{0,3}$.

For $\Phi \in \mathcal{S}$ one can define a coarsely-Lipschitz map $\theta_\Phi : \mathcal{M}(\Sigma) \to \mathcal{C}(\Phi)$. This definition is due to Masur and Minsky. We realise Φ and α to minimise the number of components of their intersection. For $a \in \mathcal{M}$ pick $\alpha \in a$ and let $\delta \subset \alpha$ be a component of $\alpha \cap \Phi$. Choose a curve γ in Φ with $\gamma \cap \delta = \emptyset$. Then let $\theta_\Phi(a) = \gamma$; this is well defined up to a bounded distance in $\mathcal{C}$. (Note that this definition assumes that Φ is not an annulus. In the case of an annulus a different definition is required. In this case, "$\mathcal{C}(\Phi)$" is a space quasi-isometric to the real line — in particular, hyperbolic. Only the logical structure is relevant here, so we will not elaborate on this point.) For a and b in $\mathcal{M}(\Sigma)$, define $\sigma_\Phi(a,b)$ to be the distance in $\mathcal{C}(\Phi)$ from $\theta_\Phi(a)$ to $\theta_\Phi(b)$.

The following theorem follows from work of Masur and Minsky.

Theorem 1.5.9 *[MM2] Let ρ be the distance function in $\mathcal{M}$. Then $\rho(a,b)$ is bounded above in terms of* $\max\{\sigma_\Phi(a,b) : \Phi \in \mathcal{S}\}$.

Theorem 1.5.10 *[BM2] For all $a, b, c, \in \mathcal{M}$ there exists $d \in \mathcal{M}$ such that for any $\Phi \in \mathcal{S}$,*

$$\rho\left(\theta_\Phi d, \mu_{\mathcal{C}(\Phi)}(\theta_\Phi a, \theta_\Phi b, \theta_\Phi c)\right)$$

is bounded by some constant depending only on the topological type of Σ.

Using Theorem 1.5.9 we see that d is well-defined up to bounded distance. Write $\mu(a,b,c) = d$ to get a map $\mathcal{M}^3 \to \mathcal{M}$. Using certain properties of subsurface projection one shows that this defines a coarse median structure on $\mathcal{M}$ [Bo1].

To obtain the bound on the rank of $\mathcal{M}$, one shows that if a quasi-square (that is, the image of a 2-cube under a coarse median homomorphism) has a large projection to both Φ and Ψ in $\mathcal{S}$, then Φ and Ψ are either disjoint or equal. It then follows that M-rk$(M) \leq \xi$ since this is the maximal number of disjoint elements of $\mathcal{S}$ that one can embed in Σ.

Similar constructions can be made to work for $\mathcal{T}$ and $\mathcal{W}$. For this one uses combinatorial models for these spaces. Specifically, it was shown by Brock [Br] that $\mathcal{W}$ is quasi-isometric to the so-called pants graph, and in [Ra] and [Du1] it is shown that $\mathcal{T}$ is quasi-isometric to the augmented marking graph. We will not give definitions here; suffice it to note that this allows us to employ similar arguments of subsurface projection. The key properties of subsurface projection needed are listed in [Bo6]. (See also [BHS2].)

We also remark that these models can be used to define the maps $\mathcal{M} \to \mathcal{T} \to \mathcal{W} \to \mathcal{C}$ mentioned earlier. (The composition of these maps, $\mathcal{M} \to \mathcal{C}$, simply selects one curve from the marking of the surface.)

The separating curve graph can be included in this picture as intermediate between $\mathcal{W}$ and $\mathcal{C}$. In most cases, it is coarse median of rank 2 [Vo]. As far as I know, its quasi-isometric rigidity has not been investigated.

1.6 Asymptotic cones

The rigidity of $\mathcal{M}$ is proven using a limiting argument phrased in terms of asymptotic cones (see [vdDW, G2]). We outline some of the ingredients here. We begin by defining the asymptotic cone of a metric space.

Let I be a countable set. Let $\mathcal{P} = \mathcal{P}(I)$ be the power set of I.

Definition 1.6.1 A subset $\mathcal{F} \subset \mathcal{P}$ is an *ultrafilter* if the following hold.

1 If $A, B \in \mathcal{F}$ then $A \cap B \in \mathcal{F}$.
2 If $A \in \mathcal{F}$ and $A \subset B$ then $B \in \mathcal{F}$.
3 If $A \subset I$ then either A or $I - A$ is in $\mathcal{F}$.
4 $\varnothing \notin \mathcal{F}$.

For example, if $a \in I$ then $\{A \in \mathcal{P}(I) \mid a \in A\}$ is an ultrafilter. An ultrafilter of this form is called a *principal ultrafilter*. Zorn's lemma implies that non-principal ultrafilters always exist (provided I is infinite).

Now fix a non-principal ultrafilter $\mathcal{F}$. If $P(i)$ is a statement depending on $i \in I$, say that $P(i)$ holds *almost always* if $\{i \mid P(i)\} \in \mathcal{F}$. For example, if (X, ρ) is a metric space and (x_i) is a sequence indexed by I, write $x_i \to x \in X$ to mean that for any $\epsilon > 0$, $|x_i - x| \leq \epsilon$ almost always. With this definition, one can readily check that any bounded sequence in $\mathbb{R}$ has a limit.

Let $(X_i, \rho_i)_i$ be a sequence of metric spaces indexed by our countable set I. Let $\mathbf{X} = \prod_i X_i$; then a point $\mathbf{x} \in \mathbf{X}$ is a sequence (x_i). Fixing a point $\mathbf{a} \in \mathbf{X}$, let $\mathbf{X}^0 = \{\mathbf{x} \in \mathbf{X} \mid \rho_i(a_i, x_i) \text{ is bounded almost always}\}$. This is independent of the choice of $\mathbf{a}$. Given $\mathbf{x}$ and $\mathbf{y}$ in $\mathbf{X}^0$, let $\rho^\infty(\mathbf{x}, \mathbf{y}) = \lim \rho_i(x_i, y_i)$, noting that $\rho_i(x_i, y_i)$ is almost always bounded by the triangle inequality. Then ρ^∞ is a pseudometric on $\mathbf{X}^0$.

Write $\mathbf{x} \sim \mathbf{y}$ if $\rho^\infty(\mathbf{x}, \mathbf{y}) = 0$. Let $X^\infty = \mathbf{X}^0 / \sim$. Then ρ^∞ descends to a metric on X^∞. It is then a general fact that X^∞ is complete; this requires that I be countable but not that the X_i be complete.

We are interested in a special case of this definition. Let (X, ρ) be a metric space and let $r_i \geq 0$ be a sequence tending to infinity. Define a new metric ρ_i on X by setting $\rho_i = \rho / r_i$. Then the limit X^∞ of the sequence of spaces (X, ρ_i) is called an *asymptotic cone* of (X, ρ). In general this might depend on the choice of r_i, or the choice of ultrafilter. However, the choice will not matter to us here.

If X is a Gromov hyperbolic space then any asymptotic cone is an $\mathbb{R}$-tree. (This is again a consequence of its treelike structure.) Conversely, a geodesic metric space all of whose asymptotic cones are $\mathbb{R}$-trees is hyperbolic [G2, Dr].

Note that some types of maps between spaces induce maps between their asymptotic cones: if $\phi: X \to Y$ is a coarsely-Lipschitz map (respectively a quasi-isometric embedding), then it induces a map $\phi^\infty : X^\infty \to Y^\infty$ that is Lipschitz (respectively bi-Lipschitz to its range). This implies, for example, that if there exists a quasi-isometric embedding $\mathbb{R}^n \to X$, then there is a bi-Lipschitz embedding $\mathbb{R}^n \to X^\infty$, so X^∞ has locally compact dimension at least n. (Recall that this is the maximal dimension of any locally compact subset.)

Note that axiom (C1) of a coarse median space Λ tells us that the median operation, μ, is coarsely-Lipschitz and so gives rise to a Lipschitz operation, $\mu^\infty : (\Lambda^\infty)^3 \to \Lambda^\infty$, on its asymptotic cone. In fact, we have the following.

Theorem 1.6.2 *If* (Λ, ρ, μ) *is a coarse median space, then* $(\Lambda^\infty, \rho^\infty, \mu^\infty)$ *is a locally convex topological median algebra with* $\text{M-rk}(\Lambda^\infty) \leq \text{C-rk}(\Lambda)$.

Note that Theorem 1.3.7 tells us that Λ^∞ has locally compact dimension at most $\text{M-rk}(\Lambda^\infty)$.

From this we can deduce Theorem 1.4.4, since any quasi-isometric embedding of H^n into Λ would give rise to a continuous (bi-Lipschitz) embedding of H^n into Λ^∞, and so $n \le \text{M-rk}(\Lambda^\infty) \le \text{C-rk}(\Lambda)$.

Some arguments will be made simpler if we can assume the metric to be a median metric. The following theorem will allow us to do this.

Theorem 1.6.3 *[Bo3, Bo6] If* $\text{C-rk}\,\Lambda < \infty$*, then* Λ^∞ *is bi-Lipschitz equivalent to a median metric via a median isomorphism.*

(In fact, under slightly stronger hypotheses applicable in the cases of interest to us, one can show that Λ^∞ embeds into a finite direct product of $\mathbb{R}$-trees by a bi-Lipschitz median homomorphism [Bo3].)

Note that by Theorem 1.3.8 we see that Λ^∞ is also bi-Lipschitz equivalent to a CAT(0) metric, and so in particular is contractible.

In general, asymptotic cones have a very complicated structure. However we have the following regularity theorem for median metric spaces. It is based on an analogous result of Kleiner and Leeb [KL].

Theorem 1.6.4 *[Bo6] Let M be a complete median metric space with* $\text{M-rk}(M) = n < \infty$*. Suppose that $f : \mathbb{R}^n \to M$ is a continuous injective map with closed image, where n is the rank of M. Then $f(\mathbb{R}^n)$ is cubulated.*

This means that $f(\mathbb{R}^n)$ is a locally finite union of n-dimensional ℓ^1-cubes: each is a convex subset of M isometric (and hence median isomorphic) to an ℓ^1 direct product of n real intervals. In other words, $f(\mathbb{R}^n)$ has the local structure of a cube complex. The complex might still bend along codimension-1 faces. However, this cannot happen if there are lots of other transverse subsets of this form.

Theorem 1.6.5 *[Bo6]. Suppose that M, f are as in Theorem 1.6.4, and suppose, in addition, that for any codimension-1 co-ordinate subspace $P \subset \mathbb{R}^n$, there is another proper embedding $f' : \mathbb{R}^n \to M$ such that $f(P) = f(\mathbb{R}^n) \cap f'(\mathbb{R}^n)$. Then $f(\mathbb{R}^n)$ is convex in M and f is a median homomorphism.*

Here a "codimension-1 co-ordinate subspace" of $\mathbb{R}^n$ is a subset of the form $\{(x_1, \dots, x_n) \mid x_i = t\}$ for some $i \in \{1, \dots, n\}$ and $t \in \mathbb{R}$.

Now let Σ be a compact surface with $\xi(\Sigma) \ge 2$. We consider the case where $\Lambda = \mathcal{M} = \mathcal{M}(\Sigma)$. By Theorem 1.6.3, $\mathcal{M}^\infty$ is bi-Lipschitz equivalent to a median metric, and so Theorems 1.6.4 and 1.6.5 apply with $n = \xi$.

Suppose that α is a pants decomposition, i.e., a collection of ξ disjoint curves on a surface of complexity ξ. Let $T(\alpha) = \{a \in \mathcal{M} \mid \alpha \subset a\}$ be the "Dehn twist flat". (Note that it is a bounded Hausdorff distance from a $\mathbb{Z}^\xi$-orbit, where $\mathbb{Z}^\xi$ is the subgroup generated by Dehn twists about the component curves.) Then the inclusion of $T(\alpha)$ into $\mathcal{M}$ induces an inclusion $T(\alpha)^\infty \subset \mathcal{M}^\infty$. In fact, we get a map $f : \mathbb{R}^\xi \to \mathcal{M}^\infty$ as in Theorem 1.6.4 with $f(\mathbb{R}^\xi) = T(\alpha)^\infty$. It turns out that it also satisfies the hypotheses of Theorem 1.6.5. The basic idea behind this is that one could replace any element $\gamma \in \alpha$ by a different curve γ' so as to give a new pants decomposition, α'. Now $T(\alpha)$ and $T(\alpha')$ remain close near a $\mathbb{Z}^{\xi-1}$-orbit (generated by Dehn twists about the components of $\alpha - \gamma = \alpha' - \gamma'$), and they diverge elsewhere. It then follows that $T(\alpha)^\infty$ meets $T(\alpha')^\infty$ in the f-image of a co-ordinate plane. Elaborating on this idea, one can verify the hypotheses of Theorem 1.6.5.

In fact, we have a converse, which we state informally as follows.

Let $f : \mathbb{R}^\xi \to \mathcal{M}^\infty$ be as in Theorem 1.6.5 (with $n = \xi$ and $M = \mathcal{M}^\infty$).

Theorem 1.6.6 *[Bo6] Sets of the form $f(\mathbb{R}^\xi) \subseteq \mathcal{M}^\infty$ (i.e., as in the hypotheses if Theorem 1.6.5) are precisely the asymptotic Dehn twist flats.*

An example of an "asymptotic Dehn twist flat" a set of the form $T(\alpha)^\infty$ as described above. However, we also need to allow sets constructed by taking I-sequences of pants decompositions rather than just a fixed pants decomposition. The key point here is that we can recognise such sets just in terms of the topology of $\mathcal{M}^\infty$.

Suppose now that $\phi : \mathcal{M} \to \mathcal{M}$ is a quasi-isometry. This induces a (bi-Lipschitz) homeomorphism $\phi^\infty : \mathcal{M}^\infty \to \mathcal{M}^\infty$. By Theorem 1.6.6, we see that ϕ^∞ preserves the collection of asymptotic Dehn twist flats. From this one can go back and deduce that ϕ sends any Dehn twist flat to within a bounded Hausdorff distance of another Dehn twist flat. Now the coarse arrangement of Dehn twist flats in $\mathcal{M}$ can be encoded in terms of the curve graph $\mathcal{C}$. It follows that ϕ gives rise to an automorphism of $\mathcal{C}$. By the combinatorial rigidity result of [I, L, K], this is induced by an element of $\mathrm{Map}(\Sigma)$, which, without loss of generality, we can take to be the identity. We now know that ϕ moves each Dehn twist flat a bounded Hausdorff distance. There are plenty of Dehn twist flats, and it follows easily that ϕ moves each point of $\mathcal{M}$ a bounded distance. This proves the quasi-isometric rigidity of $\mathcal{M}$, as formulated in Theorem 1.5.6.

In fact, we only really need that ϕ is a quasi-isometric embedding. Then ϕ^∞ maps $\mathcal{M}^\infty$ injectively onto a closed subset, which is enough to see that every asymptotic Dehn twist flat gets sent to another such. Following the argument through, this time we get an injection of $\mathcal{C}$ to itself, and the result of [Sha] tells us that it must be an isomorphism, again induced by Map(Σ). We deduce that ϕ is a quasi-isometry, and close to an element of Map(Σ). This then proves Theorem 1.5.7.

While the details are (significantly) different, related arguments can be made to work for quasi-isometries of $\mathcal{T}$ and $\mathcal{W}$, giving the rigidity results for these spaces [Bo5, Bo7]. We remark that the rigidity of $\mathcal{T}$ is independently proven in [EMR2] using quite different arguments of coarse differentiation.

References

[BaH] H.-J. Bandelt, J. Hedlikova, *Median algebras* : Discrete Math. **45** (1983) 1–30.

[BHS1] J.A. Behrstock, M.F. Hagen, A. Sisto, *Hierarchically hyperbolic spaces I: curve complexes for cubical groups* : Geom. Topol. **21** (2017) 1731–1804.

[BHS2] J.A. Behrstock, M.F. Hagen, A. Sisto, *Hierarchically hyperbolic spaces II: Combination theorems and the distance formula.* : to appear in Pacific J. Math.

[BKMM] J.A. Behrstock, B. Kleiner, Y. Minsky, L. Mosher, *Geometry and rigidity of mapping class groups* : Geom. Topol. **16** (2012) 781–888.

[BM1] J.A. Behrstock, Y.N. Minsky, *Dimension and rank for mapping class groups* : Ann. of Math. **167** (2008) 1055–1077.

[BM2] J.A. Behrstock, Y.N. Minsky, *Centroids and the rapid decay property in mapping class groups* : J. London Math. Soc. **84** (2011) 765–784.

[Bo1] B.H. Bowditch, *Coarse median spaces and groups* : Pacific J. Math. **261** (2013) 53–93.

[Bo2] B.H. Bowditch, *Invariance of coarse median spaces under relative hyperbolicity* : Math. Proc. Camb. Phil. Soc. **154** (2013) 85–95.

[Bo3] B.H. Bowditch, *Embedding median algebras in products of trees* : Geom. Dedicata **170** (2014) 157–176.

[Bo4] B.H. Bowditch, *Some properties of median metric spaces* : Groups, Geom. Dyn. **10** (2016) 279–317.

[Bo5] B.H. Bowditch, *Large-scale rank and rigidity of the Teichmüller metric* : J. Topol. **9** (2016) 985–1020.

[Bo6] B.H. Bowditch, *Large-scale rigidity properties of the mapping class groups* : Pacific J. Math. **293** (2018), no. 1, 1–73.

[Bo7] B.H. Bowditch, *Large-scale rank and rigidity of the Weil–Petersson metric* : preprint, Warwick, 2015.

[Bo8] B.H. Bowditch, *Median and injective metric spaces* : to appear in Math. Proc. Cam. Phil. Soc.

[BrH] M. Bridson, A. Haefliger, *Metric spaces of non-positive curvature* : Grundlehren der Math. Wiss. No. 319, Springer (1999).

[Br] J.F. Brock, *The Weil–Petersson metric and volumes of 3-dimensional hyperbolic convex cores* : J. Amer. Math. Soc. **16** (2003) 495–535.

[Dr] C. Druţu, *Quasi-isometry invariants and asymptotic cones* :in "International Conference on Geometric and Combinatorial Methods in Group Theory and Semigroup Theory" : Internat. J. Algebra Comput. **12** (2002) 99–135.

[Du1] M.G. Durham, *The augmented marking complex of a surface* : J. London Math. Soc. **94** (2016) 933–969.

[Du2] M.G. Durham, *The asymptotic geometry of the Teichmüller metric: Dimension and rank* : preprint, 2014, posted at ARXIV:1501.00200.

[E] R. Engelking, *Dimension theory* : North Holland (1978).

[EMR1] A. Eskin, H. Masur, K. Rafi, *Large scale rank of Teichmüller space* : Duke. Math. J. **166** (2017) 1517–1572.

[EMR2] A. Eskin, H. Masur, K. Rafi, *Rigidity of Teichmüller space* : Geom. Topol. **22** (2018) 4259–4306.

[G1] M. Gromov, *Hyperbolic groups* : in "Essays in group theory" Math. Sci. Res. Inst. Publ. No. 8, Springer (1987) 75–263.

[G2] M. Gromov, *Asymptotic invariants of infinite groups* : "Geometric group theory, Vol. 2" London Math. Soc. Lecture Note Ser. No. 182, Cambridge Univ. Press (1993).

[H] U. Hamenstädt, *Geometry of the mapping class groups III: Quasi-isometric rigidity* : preprint, 2007, posted at ARXIV:0512429.

[IT] Y. Imayoshi, M. Taniguchi, *An introduction to Teichmüller spaces* : Springer Verlag (1992).

[I] N.V. Ivanov, *Automorphism of complexes of curves and of Teichmüller spaces* : Internat. Math. Res. Notices **14** (1997) 651–666.

[KR] M. Kapovich, K. Rafi, *Teichmüller space is semi-hyperbolic* : in preparation.

[KL] B. Kleiner, B. Leeb, *Rigidity of quasi-isometries for symmetric spaces and Euclidean buildings* : Inst. Hautes Études Sci. Publ. Math. **86** (1997) 115–197.

[K] M. Korkmaz, *Automorphisms of complexes of curves in punctured spheres and on punctured tori* : Topology Appl. **95** (1999) 85–111.

[L] F. Luo, *Automorphisms of the complex of curves* : Topology **39** (2000) 283–298.

[MM1] H.A. Masur, Y.N. Minsky, *Geometry of the complex of curves I: hyperbolicity* : Invent. Math. **138** (1999) 103–149.

[MM2] H.A. Masur, Y.N. Minsky, *Geometry of the complex of curves II: hierarchical structure* : Geom. Funct. Anal. **10** (2000) 902–974.

[Mi] B. Miesch, *Injective metric on median metric spaces* : in preparation.

[Mo] L. Mosher, *Mapping class groups are automatic* : Ann. Math. **142** (1995) 303–384.

[NWZ] G.A. Niblo, N. Wright, J. Zhang, *A four point characterisation for coarse median spaces* : to appear in Groups. Geom. Dyn.

[Ra] K. Rafi, *A combinatorial model for the Teichmüller metric* : Geom. Funct. Anal. **17** (2007) 936–959.

[RS] K. Rafi, S. Schleimer, *Curve complexes are rigid* : Duke Math J. **158** (2011) 225–246.

[Ro] M.A. Roller, *Poc-sets, median algebras and group actions, an extended study of Dunwoody's construction and Sageev's theorem* : Habilitationschrift, Regensberg, 1998.

[Sha] K.J. Shackleton, *Combinatorial rigidity in curve complexes and mapping class groups* : Pacific J. Math. **230** (2007) 217–232.

[Sho] M. Sholander, *Medians and betweenness* : Proc. Amer. Math. Soc. **5** (1954) 801–807.

[vdDW] L. van den Dries, A.J. Wilkie, *On Gromov's theorem concerning groups of polynomial growth and elementary logic* : J. Algebra **89** (1984) 349–374.

[Ve] E.R. Verheul, *Multimedians in metric and normed spaces* : CWI Tract, 91, Stichting Mathematisch Centrum, Centrum voor Wiskunde en Informatica, Amsterdam (1993).

[Vo] K.M. Vokes, *Hierarchical hyperbolicity of graphs of multicurves* : posted at ARXIV:1711.03080.

[W] S. Wolpert, *Noncompleteness of the Weil–Petersson metric for Teichmüller space* : Pacific J. Math. **61** (1975) 573–577.

2

Semihyperbolicity

Martin R. Bridson
Mathematical Institute
Andrew Wiles Building
University of Oxford
Oxford OX2 6GG
European Union

Abstract

These notes on semihyperbolicity in group theory are based on a series of lectures given at the meeting *Beyond Hyperbolicity* at Cambridge in June 2016.

Introduction

Since the late 1980s, when geometric group theory emerged as a subject with a distinct identity, its mainstream has accommodated two complementary approaches to illuminating the nature of finitely generated groups. The first approach is the more classical: one gains insight by examining and constructing actions of groups on spaces with prescribed geometric properties; the quality of the insights that one obtains varies with the quality of the action – the urge to relax constraints in order to admit large classes of examples competes with the desire to focus on actions of greater quality that will illuminate groups of exceptional character and interest. The second approach came to the fore in the work of Misha Gromov [Gr1, Gr2] but was foreshadowed by the work of Max Dehn [Deh] at the beginning of the twentieth century: one regards the group itself as a geometric object, endowing it with the word metric associated to a choice of finite generating set and exploring its coarse geometry.

These two approaches are linked by two basic facts: first, if a group acts properly by isometries on a geodesic space and the quotient is compact, then the group is quasi-isometric to the space – the Švarc–Milnor Lemma [BHa1, p.140]; and second, every finitely presented group is the fundamental group of a compact 2-complex and of a closed n-manifold for each $n \geq 4$. This second basic fact provides us with proper cocompact actions of all finitely presented groups Γ on reasonable geodesic spaces, while a combination of the two facts tells us that Γ is quasi-isometric to a simply-connected manifold of bounded geometry.

These basic existence results beg natural questions about how additional features of the action might discriminate between groups – in particular they provide Γ-spaces that we might modify and refine in order to impose geometric features that will translate into large-scale geometric features of the groups acting, and thence into distinguishing algebraic properties of the groups. Manifestations of negative and non-positive curvature have commanded exceptional attention in this regard, for intrinsic reasons that will emerge from the discussion in the next section.

In recent years, the first, more-classical, approach to geometric group theory has been psychologically dominant but the second approach is always close at hand. This is particularly true of the quest to encapsulate the key features of non-uniform lattices in rank-1 Lie groups into definitions that are tight enough to ensure many powerful properties but relaxed enough to admit sought-after examples such as mapping class groups and automorphism groups of free groups. The result of this successful quest is the theory of acylindrically hyperbolic and relatively hyperbolic groups, as described by Denis Osin in [Os2] and elsewhere in this volume. This theory responds to aspects of Gromov's call for a theory of semihyperbolicity in group theory, which he articulated in [Gr1] and elaborated on in [Gr2]. In these lectures, though, I want to focus on an alternative, earlier, response to Gromov's call, which is grounded firmly in the second approach to geometric group theory: one seeks conditions on the geometry of the group itself that encapsulate the key large-scale features of non-positive curvature – i.e. a theory of *semihyperbolic* groups. I shall draw out some of the successes of this approach and describe the different classes of groups that emerged from it. I shall explain how it impinges naturally on questions of logical and linguistic complexity, relate it to more recent developments, and highlight some of the major problems in the area that remain unresolved.

There are no new results in these lectures. The main lines of the theory described were established by the mid 1990s and all of the major results were settled by the time of my ICM talk in 2006 [Bri1]. Many of the concepts have their roots in the seminal work of Gromov [Gr1, Gr2], but an important complementary perspective comes from the theory of automatic groups developed by Epstein *et al.* [ECHLPT] (also [GS]). My work with Juan Alonso [AB] established a well-defined theory of semihyperbolic groups.

I shall assume that the reader is familiar with the basic theory of CAT(0) spaces and hyperbolic groups, as described in my book with Haefliger [BHa1]. References for more specialised material are provided throughout the text.

The tone of these notes will be informal, many standard definitions will be omitted, and no detailed proofs will be given. The story is the thing.

Acknowledgements. I thank Benjamin Barrett for his excellent notes of the lectures on which this article is based. I thank Mark Hagen, Richard Webb and, above all, Henry Wilton for their assiduous, encouraging editing and helpful mathematical comments.

2.1 The universe of finitely presented groups

Consider the map of the universe of finitely presented groups sketched in Figure 2.1. This map is taken from [Bri1], to which we direct the reader for definitions and more detailed explanations.

The (large and fascinating) point labelled **1** represents the finite groups. Along the upper coast of the universe, the commutativity and polynomial growth of abelian groups slowly gives way to larger classes of amenable groups; these are separated from most of the rest of the universe by the dashed von Neumann–Tits line, which divides the groups that contain non-abelian free subgroups from those that do not. In these lectures, we focus on the regions along the bottom coast, where free groups, with their tree-like Cayley graphs, give way to the strict negative curvature of hyperbolic groups, and beyond that various classes of groups defined by increasingly weak forms of non-positive curvature.

In contrast to the other amenable groups, virtually abelian groups are indisputably non-positively curved (semihyperbolic), since they act properly and cocompactly by isometries on Euclidean space. A rich aspect of geometric group theory begins with the exploration of natural classes of semihyperbolic groups that envelop both free and free-abelian groups, interpolating between them. Foremost among these classes are the right-angled Artin groups (RAAGs), groups that act properly and cocompactly on CAT(0) cube complexes, limit groups (marked $\mathcal{L}$ on the

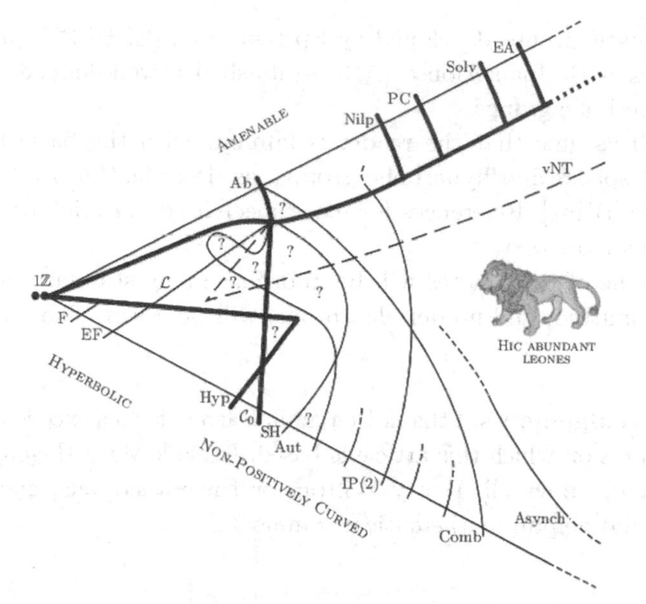

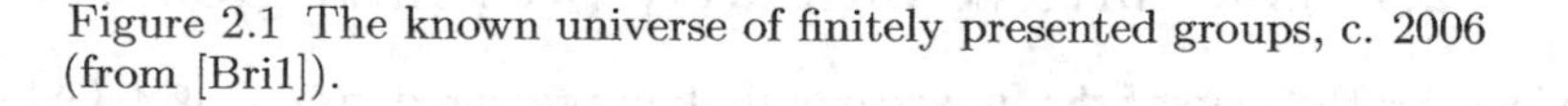

Figure 2.1 The known universe of finitely presented groups, c. 2006 (from [Bri1]).

map) and, encompassing all of these classes, the groups that act properly and cocompactly by isometries on CAT(0) spaces (marked $\mathcal{C}_0$), which are now commonly referred to as CAT(0) groups.

In his seminal essay *Hyperbolic Groups* [Gr1], Gromov defined *semi-hyperbolic groups* by the sentiment that they are the groups that "*look like* they admit a discrete cocompact action [by isometries on a CAT(0) space]." That is the theme that I want to expand upon here. Gromov's theory of hyperbolic groups will serve as a model for turning the guiding sentiment for semihyperbolic groups into a substantial theory.

Guiding Challenge 2.1.1 *Find group-theoretic definitions that capture significant aspects of the essence of negative and non-positive curvature.*

- *Seek the weakest natural conditions under which desirable theorems hold.*
- *Identify telling examples: groups of classical interest and groups constructed to distinguish between different classes.*

We are particularly interested in where the *gems of group theory* lie within the emergent theory: 3-manifold groups, mapping class groups, $\mathrm{Out}(F_n)$, and lattices in connected Lie groups, for example $\mathrm{SL}(n, \mathbb{Z})$.

Given a class of groups of interest $\mathcal{G}$, we are also interested in the finitely presented groups that arise as *subgroups* of some group in $\mathcal{G}$. Such subgroups constitute a well behaved class if $\mathcal{G}$ is to be found on the upper branch of our universe, since taking subgroups will preserve $\mathcal{G}$, but in the regions where non-positive curvature prevails, subgroups can be much wilder. In terms of actions, this means that if one weakens the standard hypothesis "properly with compact quotient" to "properly by semisimple isometries", one obtains vastly larger classes of groups.

The main classes of groups that we shall discuss are the following:

1 *Hyperbolic groups*: groups that act properly and cocompactly by isometries on δ-hyperbolic geodesic metric spaces.

2 CAT(0) *groups*: groups that act properly and cocompactly by isometries on complete CAT(0) spaces.

3 *Automatic groups*: originating in conversations of Jim Cannon and Bill Thurston on algorithmic properties of Kleinian groups, the coarse geometry of these groups resembles that of CAT(0) groups, but they are defined by their amenability to effective computation on elementary machines.

4 *Semihyperbolic groups*: to be defined, in line with Gromov's sentiment, by a definition that balances the quest for rich structure with the desire for large classes of examples.

5 *Combable groups*: marked Comb on the map, this class encompasses all of those above. These groups retain a semblance of the most characteristic feature of non-positive curvature – the convexity of the metric, as portrayed in Figures 2.2 and 2.4. Relaxations on the type of coarse paths used to capture this behaviour lead to gradations in the class of combable groups, and a central challenge is to determine which algebraic consequences of stricter forms of non-positive curvature break down at different stages of relaxation.

Plan: We shall examine the key properties of hyperbolic and CAT(0) groups and distil the features that facilitate the relevant proofs into definitions that allow us to extrapolate to larger classes of groups, exploring the limits where key theorems fail and keeping an eye on the extent to which we have included the gems of group theory.

2.1.1 The geometry of geodesics

The geometry of geodesics is central to the study of negative and non-positive curvature. In particular, many features of CAT(0) geometry can be attributed to the convexity of the metric: there is a unique geodesic arc joining any pair of points in a CAT(0) space X, and if $c_1, c_2 : [0,1] \to X$ are geodesics (parameterised proportional to arc length) then for all $t \in [0,1]$

$$d(c_1(t), c_2(t)) \leq (1-t)\, d(c_1(0), c_2(0)) + t\, d(c_1(1), c_2(1)).$$

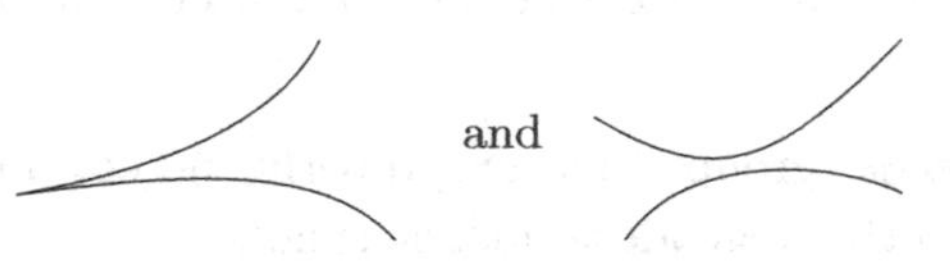

Figure 2.2 Geodesics in CAT(0) spaces

2.2 Some key features of hyperbolic groups

Gromov's theory of hyperbolic groups will serve us as a guide and a source of challenging questions. A hallmark of this theory is that diverse attempts to encapsulate the essence of negative curvature in a robust (quasi-isometry invariant) condition all define the same class of groups. The most intuitive of these conditions is that geodesic triangles in the Cayley graph of the group should be uniformly thin.

2.2.1 Slim triangles

A geodesic space X is *hyperbolic* if there is a constant $\delta > 0$ such that all geodesic triangles in the Cayley graph of X are δ-slim in the sense that each side is contained in a δ-neighbourhood of the union of the other two sides.

By definition, a finitely generated group Γ is hyperbolic if and only if its Cayley graph is a hyperbolic metric space. Random finitely presented groups (in a suitably quantified sense) are hyperbolic – there are many ways of making this precise, e.g. [Ol1].

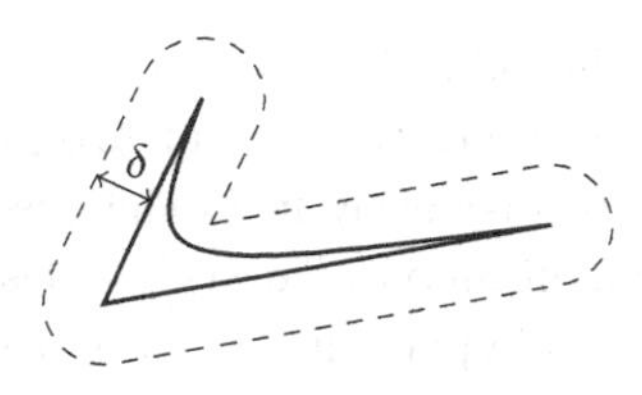

Figure 2.3 The slim triangles condition.

2.2.2 Quasi-isometries, quasi-geodesics and the Morse Lemma

The following Morse Lemma (see [BHa1, p.401]) tells us that quasi-geodesic triangles in hyperbolic spaces are slim. It follows immediately that hyperbolicity is an invariant of quasi-isometry among geodesic spaces; in particular, the hyperbolicity of a finitely generated group Γ does not depend on the choice of generators; and if Γ acts properly and cocompactly on a geodesic space X, then Γ is hyperbolic if and only if X is.

Recall that a (λ, ε)–quasi-isometric embedding of metric spaces is a map $f : Y \to X$ such that

$$\frac{1}{\lambda} d(y, y') - \varepsilon \leq d(f(y), f(y')) \leq \lambda\, d(y, y') + \varepsilon$$

for all $y, y' \in Y$. If Y is an interval of the real line then f (or its image) is called a (λ, ε)-*quasi-geodesic.* If there is a quasi-isometric embedding $g : X \to Y$ such that $f \circ g$ and $g \circ f$ are uniformly close to the identity, then f is called a *quasi-isometry* (with quasi-inverse g).

Proposition 2.2.1 (Morse Lemma) *Fix $\lambda \geq 1$ and $\epsilon \geq 0$. If X is a δ-hyperbolic geodesic space, then every (λ, ϵ)-quasi-geodesic in X is contained in the K-neighbourhood of a geodesic with the same endpoints, where K depends only on δ, λ and ϵ.*

The Morse Lemma carries much of the essence of negative curvature and it fails badly when one moves to spaces of non-positive curvature: in the Euclidean plane $\mathbb{E}^2$, for example, the logarithmic spiral $\sigma(t) = te^{i \log(t+1)}$ is a quasi-geodesic but it is not uniformly close to any geodesic and never settles on an asymptotic direction. This last phrase points to the fact that whereas it is natural to construct a visual sphere at infinity out of quasi-geodesic rays in a hyperbolic group, one cannot hope to do the same in a semihyperbolic setting – this weakness in the *geometry at infinity* is a recurring thorn in the side of all versions of coarse semihyperbolic geometry; see [FO] for an interesting recent attempt to resolve this.

The important lesson to take away from this discussion is that, implicitly, hyperbolicity is a condition on *all* quasi-geodesics in a space whereas this cannot be true of semihyperbolicity, so more **choices** will have to be made.

2.2.3 Local recognition of quasi-geodesics

Related to the Morse Lemma, in hyperbolic spaces we have *local recognition of efficient quasi-geodesics.* (We shall address the question of efficiently recognising geodesics in Section 2.2.9.)

A *k-local-geodesic* is a path whose restriction to each subinterval of length at most k is a geodesic.

Lemma 2.2.2 *There exist functions $\alpha = \alpha(\delta)$ and $\beta = \beta(\delta)$ such that if X is a δ-hyperbolic geodesic space, then every $(8\delta + 1)$-local-geodesic is an (α, β)-quasi-geodesic.*

A crucial feature underpinning many results about CAT(0) spaces is that local-geodesics are global geodesics in such spaces, but this fails completely when one quasifies the situation: local recognition is lost. For example, in the Cayley graph of $\mathbb{Z}^2 = \langle x, y \mid [x, y] \rangle$, the (square) loop labelled $x^k y^k x^{-k} y^{-k}$ is a k-local-geodesic, but it is not a (λ, ε)-quasi-geodesic when $k > \varepsilon\lambda/4$.

2.2.4 Algorithmic construction of finite classifying spaces

If Γ is hyperbolic then it will act properly and cellularly on a contractible simplicial complex that can be constructed algorithmically from a finite presentation of Γ. If Γ is torsion-free then the quotient of the action is a finite $K(\Gamma, 1)$; in general, the stabilisers of simplices will be finite [BHa1, p.468] and the fixed-point sets of finite subgroups will be contractible [MS].

Explicitly, from a finite presentation one can compute δ such that Γ is δ-hyperbolic [P1], and if $R > 4\delta + 4$ then the *Rips Complex* $P_R(\Gamma)$ is contractible, where P_R is the simplicial complex with a k-simplex $[x_0, \ldots, x_k]$ for each subset $\{x_0, \ldots, x_k\}$ of Γ with $d(x_i, x_j) < R$ for all i and j. The action of Γ on $P_R(\Gamma)$ is by left multiplication.

The contractibility of $P_R(\Gamma)$ is a consequence of the contracting properties of the *geodesic combing* in a hyperbolic space: if one pushes the most remote point of a finite set of diameter less than R a distance $R/2$ towards a basepoint along any choice of geodesic, then the diameter of the modified set will remain less than R; see [BHa1, p.469]. This property is characteristic of strictly negative curvature: it fails in a Cayley graph of $\mathbb{Z}^n$ for example. Nevertheless, we shall see in Section 2.7 that combings in wider classes of groups (with geodesics replaced by more selected paths exhibiting a weaker contracting property) facilitate the construction of classifying spaces in significant generality.

2.2.5 Isoperimetric inequalities and the word problem

In the real hyperbolic plane $\mathbb{H}^2$ every loop of length ℓ bounds a disc of area at most ℓ. This is the **linear isoperimetric inequality** of hyperbolic geometry; it extends (with suitable constants) to any simply connected manifold whose sectional curvatures are negative and bounded away from 0. With a suitably defined coarse notion of area [BHa1, III.H.2], it also extends to a large class of hyperbolic spaces including hyperbolic groups.

The type of isoperimetric inequality satisfied by a space that a group Γ acts on properly and cocompactly is closely related to the difficulty of the *word problem* in Γ as measured by its Dehn function; see [Bri10].

Definition 2.2.3 Let $\langle A \mid R\rangle$ be a finite presentation for a group Γ. For each $w \in F(A)$ with $w =_\Gamma 1$, define

$$\text{Area}(w) := \min\{N \mid w =_{F(A)} \prod_{i=1}^{N} \theta_i r_i \theta_i^{-1} \text{ with } r_i \in R^\pm\}.$$

The *Dehn function* $\Delta\colon \mathbb{N} \to \mathbb{N}$ for Γ is

$$\Delta(n) = \max\{\text{Area}(w) \mid |w| \le n,\ w =_\Gamma 1\}.$$

The Dehn functions associated to different finite presentations of the same group (or anything quasi-isometric to it) are equivalent $f \simeq g$, where $\simeq$ is the symmetrisation of the following relation: $g \preceq f$ if there is a constant $C > 0$ such that $g(n) \le C\, f(Cn) + Cn + C$. This allows us to write Δ_Γ without ambiguity, provided that we are only interested in the $\simeq$-equivalence class of Δ.

If Γ acts properly and cocompactly by isometries on a simply-connected geodesic space X of bounded curvature (e.g. a smooth manifold) then the Dehn function of Γ is $\simeq$-equivalent to the function that gives the least upper bound on the area of minimal-area discs filling loops in X (where the bound is given as a function of the length of the loop); see [Bri10].

Gromov [Gr1] proved that the linear isoperimetric inequality characterises hyperbolic groups. Careful proofs of this were given by several authors – see [BHa1, p.419] for commentary and references.

Theorem 2.2.4 *A group* Γ *is hyperbolic if and only if it is finitely presented and* $\Delta_\Gamma(n) \simeq n$.

This striking fact is accentuated by another result of Gromov: if a group satisfies a subquadratic isoperimetric inequality, then it is hyperbolic. Again, detailed proofs were given by several authors – see [Bow], [Ol2], [BHa1, p.422].

Theorem 2.2.5 *If* $\Delta_\Gamma(n) = o(n^2)$ *then* $\Delta_\Gamma(n) \simeq n$.

Brady and Bridson [BeB] subsequently proved that Gromov's gap is the only one in the *isoperimetric spectrum*

$$\mathrm{IP} = \{\alpha \mid \exists \Gamma \text{ with } \Delta_\Gamma \sim n^\alpha\}.$$

1 2 dense

It was later shown [BBFS] that IP contains all rational numbers bigger than 2. More definitive information about IP was established by Birget, Ol'shanskii, Rips and Sapir in [BORS], which was extended very recently by Ol'shanskii [Ol3].

Smooth manifolds with non-positive sectional curvature and, more generally, complete CAT(0) spaces, satisfy a quadratic isoperimetric inequality: suitable efficient filling discs can be constructed simply by coning off a loop to any point along its length using geodesics – see [BHa1, p.444]. Thus $\mathrm{IP}(2) = \{\Gamma \mid \Delta_\Gamma(n) \simeq n^2\}$ appears as a potentially reasonable candidate for the class of semihyperbolic groups. But after long investigation, we now know that IP(2) is much too wild a zoo to merit this designation: its exact composition remains a mystery, but it certainly contains groups that do not share key characteristics of non-positive curvature – see, for example, [DERY], [Gu], [Y1], [OlS].

2.2.6 Other decision problems

Theorem 2.2.6 *In hyperbolic groups there is a rapid solution to the conjugacy problem (linear time in a RAM model of computation), and* *an equally efficient algorithm to determine conjugacy among finite lists of elements.*

The following is a more precise statement of this fact. Here, for brevity, we write $|\gamma|$ in place of $d(1,\gamma)$ for the length of a shortest word in the generators that represents γ.

If Γ *is a* k*-generated,* δ*-hyperbolic group, and the lists* $(a_1,\ldots,a_m)$ *and* $(b_1,\ldots,b_m)$ *are conjugate in* Γ*, then there exists* $\gamma\in\Gamma$ *such that* $a_i^{\gamma}=b_i$ *for* $i=1,\ldots,m$ *and* $|\gamma|<C\max_{i,j}\{|a_i|,|b_j|\}$*, where* C *is a constant that depends only on* δ,k *and* m*. Moreover, there is an algorithm that will find* γ *in linear time (with a RAM model of computation).*

See [BHa1, p.453], [EH], [BHa2] and [BH] for proofs but note that the solvability of the conjugacy problem was already known to Gromov [Gr1].

Sketch proof for single elements: In a free group, each element (word) has a cyclically-reduced form $\bar{w}$, and w_1 and w_2 are conjugate if and only if $\bar{w}_2$ is a cyclic permutation of $\bar{w}_1$ – something that is easily checked. Analogously, in negatively curved (locally CAT(−1)) spaces, there is a unique closed geodesic representing each conjugacy class in the fundamental group, and a naive tightening process makes a loop in the homotopy class converge rapidly to this geodesic. In a general hyperbolic group, given words u and v one can "tighten" to cyclically reduced $(8\delta+4)$-local-geodesics $\bar{u}$ and $\bar{v}$, and u and v are conjugate if and only if there exists w with $|w|\le 4\delta+4$ such that $w^{-1}\bar{u}w=\bar{v}$. □

Question 2.2.7 How difficult is it to determine conjugacy in the setting of non-positive curvature? To what extent can we hope to mimic the above proof?

The following theorem is one of the deepest concerning hyperbolic groups. It was proved by Sela [Se] in the torsion-free rigid case. The rigidity hypothesis was removed by Dahmani and Groves [DGr], and Dahmani and Guirardel [DGu] proved the theorem in full generality.

Theorem 2.2.8 *The isomorphism problem is solvable in the class of hyperbolic groups.*

Again, we would like to extend this to (large classes of) semihyperbolic groups. Added interest in the isomorphism problem for such groups comes from the close connection with the recognition problem for high-dimensional manifolds, which is established by the following celebrated theorem of Farrell and Jones [FJ2].

Theorem 2.2.9 *If $n \geq 5$ and M^n and N^n are closed non-positively curved manifolds of dimension n, then M is homeomorphic to N if and only if $\pi_1 M \cong \pi_1 N$.*

In the light of Sela's theorem, we deduce:

Corollary 2.2.10 *If $n \geq 5$, the homeomorphism problem is solvable among closed n-dimensional manifolds that support a metric of negative curvature.*

One would be able to extend this corollary to the non-positively curved case if one could solve the isomorphism problem among the fundamental groups of closed non-positively curved manifolds (or a larger class of semihyperbolic groups), but this problem remains open.

2.2.7 Abelian subgroups and translation numbers

In a hyperbolic group, the cyclic subgroup generated by any element γ of infinite order has finite index in its centralizer $C_\Gamma(\gamma)$, so in particular all abelian subgroups are virtually cyclic. Since $\mathbb{Z}^n$ is a prototypical semihyperbolic group, this control cannot extend to the semihyperbolic setting, but one might hope that all abelian subgroups would be finitely generated (excluding subgroups such as $\mathbb{Z}[1/2]$ or $\mathbb{Q}$ for example).

One might also ask about the *geometry* of abelian subgroups. In a hyperbolic group, if γ has infinite order then $n \mapsto \gamma^n$ is a quasi-geodesic. In what generality might one prove that abelian subgroups of semihyperbolic groups are quasi-isometrically embedded?

More subtle structure emerges when one examines the set of *translation lengths*

$$\tau(\gamma) := \lim_{n\to\infty} d(1, \gamma^n)/n.$$

A remarkable theorem of Gromov, given an elegant proof by Delzant [Del] (cf. [BHa1, p.466]), shows that in the hyperbolic setting translation lengths are positive rational numbers with bounded denominators.

Theorem 2.2.11 *If Γ is hyperbolic, then there exists an integer $N \in \mathbb{N}$ such for all elements $\gamma \in \Gamma$ of infinite order, $\tau(\gamma)N$ is a positive integer.*

2.2.8 Torsion elements

Theorem 2.2.12 *Each hyperbolic group has only finitely many conjugacy classes of finite subgroups.*

Sketch of proof: This is a coarse version of the standard proof of the Cartan fixed point theorem from CAT(0) geometry. The set of coarse circumcentres for any bounded set of points in a hyperbolic space is of uniformly bounded diameter (see [BHa1, p.459]); when applied to the elements in a finite subgroup, this establishes the desired finiteness result.

2.2.9 Markov properties: recognition of geodesics

The theory of *automatic groups* began with Jim Cannon's insight [C] that there is a remarkably simple algorithm (looking at balls of bounded radius) for recognising geodesics in a hyperbolic group. The following theorem will be explained in more detail in Section 2.5.

Theorem 2.2.13 *If Γ is a hyperbolic group with finite generating set $A = A^{-1}$ then there is a finite state automaton M with input alphabet A so that the language of words accepted by M is exactly the set of words labelling geodesics in the Cayley graph* $\mathrm{Cay}(\Gamma, \mathrm{A})$.

Standard arguments allow one to deduce from this that functions counting geodesics in hyperbolic groups are remarkably simple.

Corollary 2.2.14 *If s_n is the number of elements γ of Γ with $d(1,\gamma) = n$, then $\zeta_{\Gamma,A}(t) := \sum_{n=0}^{\infty} s_n t^n$ is a rational function.*

To what extent might this phenomenon persist in non-positive curvature?

2.3 Some properties of CAT(0) groups

The following list of properties is taken from [BHa1, pp.438-448]; it should be compared with the results in the previous section. This list provides us with a list of desirable properties as we attempt to frame a theory of semihyperbolic groups.

Theorem 2.3.1 *If Γ acts properly and cocompactly by isometries on a complete* CAT(0) *space X, then it has the following properties.*

1 *Γ is finitely presented.*
2 *If Γ is torsion-free then it has a finite $K(\Gamma,1)$.*
3 *Γ has only finitely many conjugacy classes of finite subgroups.*
4 *If $\gamma \in \Gamma$ has infinite order, then the centraliser $C_\Gamma(\gamma)$ has a subgroup of finite index that splits as a direct product $\langle\gamma\rangle \times D_\gamma$.*

5 $\{\tau(\gamma) : \gamma \in \Gamma\}$ *is discrete and* $\tau(\gamma) > 0$ *if* γ *has infinite order.*

6 THE FLAT TORUS THEOREM: *If* $A < \Gamma$ *is virtually abelian of rank* n*, then there is an* A*-invariant isometric embedding* $\mathbb{E}^n \hookrightarrow X$ *on which* A *acts cocompactly.*

7 *Ascending chains* $A_0 < A_1 < \cdots < \Gamma$ *of virtually abelian subgroups are finite.*

8 THE SOLVABLE SUBGROUP THEOREM: *Every virtually solvable subgroup* $S < \Gamma$ *is finitely generated and virtually abelian.*

9 WORD PROBLEM: *The Dehn function* $\Delta_\Gamma(n)$ *is* $\simeq n$ *or* $\simeq n^2$*.*

10 CONJUGACY PROBLEM: *For each finite generating set* A *of* Γ*, there are constants* C, λ *such that if* γ, γ' *are conjugate in* Γ *and* $m = \max\{|\gamma|, |\gamma'|\}$*, then there is a sequence* $\gamma = \gamma_1, \ldots, \gamma_n = \gamma'$ *where* $\gamma_{i+1} = a_i \gamma_i a_i^{-1}$ *with* $a_i \in A^{\pm 1}$*, all* $|\gamma_i| \leq \lambda \max\{|\gamma|, |\gamma'|\}$*, and* $n \leq C^m$*.*

In many circumstances one can reduce the exponential bound C^m in (10) to a linear bound, as in Theorem 2.2.6, but it is unknown if one can do so in general.

Finite presentability and the nature of Dehn functions are coarse invariants that will transfer easily to semihyperbolic groups. This is also true of the higher finiteness properties "*type* F_n", which record the existence of a $K(\Gamma, 1)$ with finite n-skeleton. On the other hand, the solubility of conjugacy problems is far from a coarse invariant – there exist finitely presented groups $H < G$ with $[G : H] = 2$ such that G has a solvable conjugacy problem and H does not [CM] – so transferring (10) to the semihyperbolic setting requires more care. The other properties listed all rely on more subtle, local properties of CAT(0) geometry, in particular product decomposition theorems that are destroyed when one perturbs the local geometry. Thus expecting them to transfer to a coarser, semihyperbolic setting is less reasonable, and one should at least be prepared to accept that the semihyperbolic shadows of these properties will be considerably weaker.

2.4 Combings and semihyperbolicity

Having drawn motivation and orientation from the prototypes of CAT(0) and hyperbolic groups, it is time to formulate definitions for non-positively curved groups, phrased in terms of the intrinsic geometry of the group. We focus on the geometry of geodesics in a CAT(0) space X as portrayed in Figure 2.2. We consider a group Γ acting properly and cocompactly by isometries on X, fix a basepoint $p \in X$ and pull the geodesic segments $[p, \gamma.p]$ back to Γ by means of the Γ-equivariant quasi-isometry $\Gamma \to X$ defined by $\gamma \mapsto \gamma.p$.

The resulting paths, which we denote by σ_γ, will be (λ, ε)-quasi-geodesics in Γ, where the uniform constants λ, ε depend on the action and the chosen generating set A (word metric) on Γ. It is an easy matter to approximate these uniformly by edge-paths in the Cayley graph of Γ, which can aid intuition. It is also helpful to identify edge-paths in the Cayley graph that start at the identity with the words in the generators that label these edge-paths. Thus we regard the choice of quasi-geodesics σ_γ as a section (right-inverse)

$$\sigma : \gamma \mapsto \sigma_\gamma$$

to the natural map to Γ from the free monoid (i.e. set of finite words) $(A \sqcup A^{-1})^*$.

Following Bill Thurston, we define any such section σ to be a COMBING of Γ. (In settings with a more algebraic emphasis, this would be called a normal form; cf. Section 2.5.)

The convexity of the metric on X (figure 2.2) implies that for some constant $k > 0$ and all $\gamma, \gamma' \in \Gamma$ we have

$$d(\sigma_\gamma(t), \sigma_{\gamma'}(t)) \leq k\, d(\gamma, \gamma') \tag{2.1}$$

at all integer times t. (By definition, $\sigma_\gamma(t)$ is the image in Γ of the prefix of length t in the word σ_γ; for times $t \geq |\sigma_\gamma|$ the path is stationary at γ.) This is called the (1-sided) FELLOW-TRAVELLER PROPERTY, with k the fellow-traveller constant.

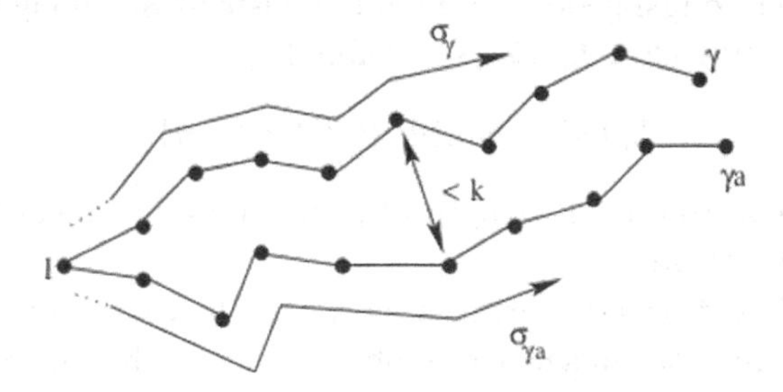

Figure 2.4 The fellow-traveller property

A combing $\sigma : \Gamma \to (A \sqcup A^{-1})^*$ defines a choice of preferred path for any pair of elements of Γ: the path joining γ_1 to γ_2 is $\gamma_1.\sigma_{\gamma_1^{-1}\gamma_2}$.

The convexity of the function $t \mapsto d(c_1(t), c_2(t))$ for geodesics in a

CAT(0) space follows easily from the special case $c_1(0) = c_2(0)$, but the analogous statement for combings is false: the fellow-traveller property does not imply the 2-SIDED FELLOW-TRAVELLER PROPERTY

$$d(g.\sigma_{g^{-1}\gamma'}(t),\ \sigma_\gamma(t)) \le k\ (d(1,g) + d(\gamma,\gamma')) \tag{2.2}$$

for all $g, \gamma, \gamma' \in \Gamma$.

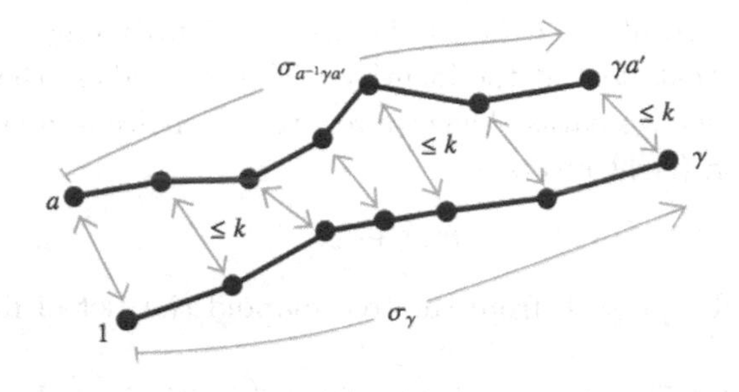

Figure 2.5 The 2-sided fellow-traveller property

The combings in our prototype – i.e., those associated to actions on CAT(0) spaces – do enjoy this stronger property, so it might be that we need to require it in our theory of semihyperbolic groups in order to obtain the theorems that we seek; or maybe not! Maybe the classes of groups defined by the 1-sided and 2-sided conditions are the same, or at least have essentially the same properties; or maybe not!

Another dilemma concerns the length and the quasi-uniform speed of the paths σ_γ: does it matter that the combing lines that arise in the prototype were quasi-geodesics; might it instead be enough to control the LENGTH FUNCTION $L_\sigma : \mathbb{N} \to \mathbb{N}$ defined by

$$L_\sigma(n) = \max\{|\sigma_\gamma| : d(1,\gamma) \le n\}.$$

Or perhaps controlling the length of paths does not affect the class of groups defined at all.

A more subtle distinction arises when we ask how hard it is to recognise which words are in the image of a combing σ; this is the subject of Section 2.5.1, but we immediately ask if it is always possible to arrange for the image of σ to be recognised by a finite state automaton – i.e., for it to be a **regular language** – as is the case for geodesic combings of hyperbolic groups (Cannon's Theorem 2.2.13).

Finally, we might also question whether the fellow-traveller property is always the most appropriate notion of closeness for paths: would it make a difference if we were instead to work with the Gromov–Hausdorff distance between the images of paths, or allow monotone reparametrisation of paths and consider the ASYNCHRONOUS FELLOW-TRAVELLER PROPERTY, i.e. controlled

$$\inf_{\rho_1,\rho_2} \max\{d(\sigma_\gamma(\rho_1(t)), \sigma_{\gamma'}(\rho_2(t))) \mid t \in \mathbb{N}\} \leq k\, d(\gamma, \gamma')$$

(or the 2-sided equivalent) where the infimum is taken over proper maps $\rho_i : \mathbb{N} \to \mathbb{N}$ with $\rho_i(0) = 0$ and $\rho_i(t+1) \in \{\rho_i(t), 1 + \rho_i(t)\}$. (See [Bri3], [Ge1] and [BGSS] for more on this.)

Definition 2.4.1 [Combable, semihyperbolic, etc.]

1 A finitely generated group Γ is *combable* (resp. *bicombable*) if it has a combing σ with the fellow-traveller property (resp. 2-sided fellow-traveller property).
2 A bicombable group Γ is *semihyperbolic* if, in addition, there exist constants $\lambda > 1$ and $\varepsilon > 0$ such that the combing lines σ_γ are (λ, ε)-quasi-geodesics.
3 A combable (resp. bicombable) group is *automatic* (resp. *biautomatic*) if the language $\mathcal{L}_\sigma = \{\sigma_\gamma \mid \gamma \in \Gamma\} \subset (A \sqcup A^{-1})^*$ formed by the combing words is *regular*.
4 A finitely generated group Γ is *asynchronously combable* if it has a combing σ with the asynchronous fellow-traveller property. If $\mathcal{L}_\sigma$ is regular then Γ is *asynchronously automatic*.

Requiring $\mathcal{L}_\sigma$ to be a regular language imposes subtle constraints on the geometry of the combing. For synchronous combings, it forces the combing lines σ_γ to be quasi-geodesics with uniform constants, so biautomatic groups are semihyperbolic. But it is important to note that it is a combination of the synchronous fellow-traveller property and regularity that yields this constraint: in contrast, there exist asynchronous automatic groups that do *not* admit quasi-geodesic combings with the asynchronous fellow-traveller property – see [Bri12] for this and a detailed study of how the length function L_σ of an optimal asynchronous automatic structure can vary with the group.

2.4.1 Quasi-convex subgroups

In the setting of hyperbolic groups, if the inclusion of a finitely generated subgroup $H \hookrightarrow \Gamma$ is a quasi-isometric embedding, then H will be *quasi-convex* in the sense that there is a constant k such that every geodesic with endpoints in H lies in the k-neighbourhood of H, and H will be hyperbolic. (This is a consequence of the Morse Lemma.) If a subspace of a CAT(0) space $Y \hookrightarrow X$ is convex, in the sense that any geodesic with endpoints in Y is entirely contained in Y, then it is isometrically embedded and CAT(0) in the induced metric, but this does not extend to quasi-isometric embeddings.

In contrast to these classical situations, when one coarsens and considers semihyperbolic groups, isometrically embedded subgroups can be utterly wild – they need not be finitely presented, for example (see [BHa1, p.486]). But if one adapts the notion of quasi-convexity to fit the combing σ, then quasi-convex subgroups are seen to be the natural analogue of quasi-convex subgroups of hyperbolic groups and isometric subspaces of CAT(0) spaces.

Definition 2.4.2 Let Γ be a group with combing σ. A subgroup $H < \Gamma$ is σ-quasi-convex if there is a constant $k > 0$ such that the path σ_h lies entirely within the k-neighbourhood of H for every $h \in H$.

Proposition 2.4.3 *Let Γ be a finitely generated group with combing σ and let $H < \Gamma$ be a σ-quasi-convex subgroup. If σ satisfies the fellow-traveller property (resp. the 2-sided fellow-traveller property), then H is finitely presented and combable (resp. bicombable). If σ is a semihyperbolic structure, then H is semihyperbolic and $H \hookrightarrow \Gamma$ is a quasi-isometric embedding.*

A proof of this proposition, and all of the results in this section, can be found in [AB].

A further important observation is that for combings with the 2-sided fellow-traveller property, the intersection of a pair of quasi-convex subgroups is again quasi-convex.

Several important results in the theory of semihyperbolic groups rely on the fact that certain subgroups are *intrinsically quasi-convex* – i.e. are quasi-convex with respect to all choices of semihyperbolic structure. This insight originates in the work of Gersten and Short on biautomatic groups [GS].

Proposition 2.4.4 *If Γ is semihyperbolic with combing σ, then for all finite sets $T \subset \Gamma$, the centraliser $C_\Gamma(T)$ is σ-quasi-convex.*

In situations such as this proposition, where a subgroup C is quasi-convex with respect to all semihyperbolic structures, one says simply that C *is quasi-convex* in Γ.

2.4.2 A summary of basic properties

Our focus is mainly on the semihyperbolic groups. The following is a compilation of theorems for these groups – see the original paper [AB] or [BHa1, III.Γ.4].

Theorem 2.4.5 *If* Γ *is semihyperbolic then it has the following properties.*

1 Γ *is finitely presented.*

2 Γ *has a classifying space* $K(\Gamma, 1)$ *with finitely many cells in each dimension.*

3 The centraliser $C_\Gamma(T)$ *of every finite subset* $T \subset \Gamma$ *is quasi-convex, hence semihyperbolic.*

4 If $\gamma \in \Gamma$ *has infinite order, then*[1] $\tau(\gamma) > 0$.

5 ALGEBRAIC FLAT TORUS THEOREM: *Every finitely generated abelian subgroup* $A < \Gamma$ *is quasi-convex.*

6 THE POLYCYCLIC SUBGROUP THEOREM: *Every polycylic subgroup* $S < \Gamma$ *is virtually abelian.*

7 WORD PROBLEM: *The Dehn function* $\Delta_\Gamma(n)$ *is* $\simeq n$ *or* $\simeq n^2$.

8 CONJUGACY PROBLEM: *For each finite generating set* A *of* Γ*, there are constants* C, λ *such that if* γ, γ' *are conjugate in* Γ *and* $m = \max\{|\gamma|, |\gamma'|\}$*, then there is a sequence* $\gamma = \gamma_1, \dots, \gamma_n = \gamma'$ *where* $\gamma_{i+1} = a_i \gamma_i a_i^{-1}$ *with* $a_i \in A^{\pm 1}$*, all* $|\gamma_i| \leq \lambda \max\{|\gamma|, |\gamma'|\}$*, and* $n \leq C^m$.

This compilation should be compared with Theorem 2.3.1. The absence of virtual splittings for centralizers is explained by Example 2.6.2. The weaker statements for abelian and solvable groups is frustrating: it remains unknown if every abelian subgroup of a semihyperbolic group is finitely generated, and unknown if every solvable subgroup has to be virtually abelian. This deficit is connected to an issue concerning translation numbers: it is unknown if $\{\tau(\gamma) : \gamma \in \Gamma\}$ has to be discrete. Another deficiency in need of a proof or telling example is the fact that we do not know if a torsion-free semihyperbolic group must have a finite classifying space.

As one weakens the constraints on the geometry of σ one can prove less. The following is a compilation of results from [Al], [ECHLPT] [Bri3], [Bri12], [Ge2], [Bri6], [Bri7], [AB], [Sh].

[1] easy exercise: the statement $\tau(\gamma) > 0$ is independent of the generating set

Theorem 2.4.6

1 If Γ is asynchronously combable, with combing σ, then Γ is finitely presented and of type FP_∞*;*

2 moreover, the word problem in Γ is solvable: its Dehn function satisfies $\Delta_\Gamma(n) \preceq n\, L_\sigma(n)$; it is at most exponential, but need not be quadratic (Example 2.4.9).

3 If Γ is bicombable, then its conjugacy problem is solvable, indeed (8) of Theorem 2.4.5 holds;

4 moreover, centralisers of finite subsets $T \subset \Gamma$ are quasi-convex, hence finitely presented and bicombable.

5 There exist combable groups in which the conjugacy problem Γ is unsolvable.

6 There exist combable groups in which the centralisers of certain finite subsets $T \subset \Gamma$ are not finitely generated.

Combining groups. All of the above classes of groups are closed under the formation of free products and direct products. (This is an easy exercise.) Amalgamations and HNN extensions along finite subgroups also behave well (see [AB] for the semihyperbolic case), but one has to be very careful with more general amalgamated free products: there are positive general results under suitably strict hypotheses (see [BGSS], for example), but there are also sobering examples – the amalgamation of two free abelian groups along subgroups of index 2, for instance, need not be asynchronously automatic and can have exponential Dehn function [Bri8].

The classes of groups that are defined by 1-sided fellow-traveller conditions are closed under finite extensions (again this is easy), but those defined by 2-sided conditions are probably not (although, surprisingly, no distinguishing examples have been found).

Remark 2.4.7 If one replaces the fellow-traveller property by the asynchronous fellow-traveller property, many more groups of geometric interest are admitted – see [Bri2], for example.

The ISOMORPHISM PROBLEM remains open in the class of semihyperbolic groups and the class of automatic groups, but in the larger and wilder class of combable groups we have (see [Bri6]):

Theorem 2.4.8 *The isomorphism problem is unsolvable in the class of combable groups. More precisely, there exist recursive sequences of finite presentations (P_n), where the number of generators and relations does* *not vary with n, such that there is no algorithm to determine which of the groups presented are isomorphic to the group with presentation P_1.*

Geometric group theory in the 1990s was marred by the absence of examples to distinguish between the different classes of groups described above, but this unhappy state of affairs has been largely resolved.

Distinguishing Example 2.4.9 The group whose presentation is given below is obtained by doubling the direct product of free groups $F(a,b) \times F(c,d)$ along the subgroup $\langle c, ad\rangle$. It is combable but not bicombable or automatic. Its Dehn function is cubic and the centraliser of $\{a_1, a_2, b_1, b_2\}$ is not finitely generated (see [Bri7]).
Generators: $a_1, b_1, a_2, b_2, c, d_1, d_2$
Relations: $a_1 d_1 = a_2 d_2$, $[a_i, c] = [a_i, d_i] = [b_i, c] = [b_i, d_i] = 1$, $i = 1, 2$

Distinguishing Example 2.4.10 We describe how to construct a combable group in which the conjugacy problem is *unsolvable.* This provides a further construction of combable groups that are not bicombable, but more importantly it reaffirms the wildness of the class of combable groups indicated by Theorem 2.4.8, and provides definitive proof that this wildness is greater than that to be found among semihyperbolic and bicombable groups.

A famous construction of Rips [R] assigns to each finite presentation of a group Q a pair $N < H$ where H is a torsion-free hyperbolic group, N is a 2-generator normal subgroup, and $H/N \cong Q$. If Q has an unsolvable word problem (we might take Borisov's 5-generator, 12-relator presentation, for example [Bor]), then there is no algorithm that can determine which words in the generators of H define elements of N; we fix such a Q, which we also assume to be torsion-free.

Let F be a free group of rank 2, let $\phi : F \to N$ be an epimorphism, let $\overline{F} \subset F \times N$ be the graph of ϕ, and let $D = \Delta(F \times H; \overline{F})$ be the amalgamated free product of two copies of $F \times H$ along $\overline{F}$. It is proved in Theorem A of [Bri6] that D is combable but has an unsolvable conjugacy problem.

We close this section by pointing out that essentially *any* non-trivial geometric constraint on a combing has consequences for the group.

Remark 2.4.11 Given any finitely generated group Γ, an arbitrary choice of geodesic word representing each element will define a combing σ with *width* $\phi_\sigma(n) \le n$, where

$$\phi_\sigma(n) = \max\{d(\sigma_\gamma(t), \sigma_{\gamma'}(t)) \mid t \in \mathbb{N};\, d(1,\gamma), d(1,\gamma') \le n;\, d(\gamma,\gamma') = 1\}.$$

But if Γ admits a combing with $\phi_\sigma(n) < n$ for large n (with no constraint on the length of the combing lines σ_γ) then Γ will be finitely presented and have a solvable word problem! For more regarding this observation, see [Bri3] and subsequent references to that paper.

2.5 Languages and the complexity of normal forms

So far we have said little about the Markov properties that we saw in the context of hyperbolic groups and nothing about the problem of recognising whether a word in the generators of a non-positively curved group is (quasi-)geodesic or is the label on a combing line. That is the subject of this section.

We want to be careful about what we mean when we endow a group Γ with a generating set A: this is a choice of epimorphism $\mu\colon (A^{\pm})^* \to \Gamma$ from the free monoid on $A^{\pm}$ such that $\mu(a)^{-1} = \mu(a^{-1})$, where $A^{\pm}$ is the disjoint union of A and a set of formal symbols $\{a^{-1} : a \in A\}$. We repeat the definition of a combing, giving it a more algebraic emphasis.

Definition 2.5.1 A *normal form* or *combing* is a section $\sigma\colon \Gamma \to (A^{\pm})^*$ of μ, i.e. $\mu\sigma_\gamma = \gamma$ for all γ in Γ. Thus σ_γ is a choice of word representing $\gamma \in \Gamma$.

In the previous section we considered how weak convexity conditions on the geometry of $\mathcal{L}_\sigma = \{\sigma_\gamma \mid \gamma \in \Gamma\}$, regarded as a set of discrete paths in Γ, imposed constraints on Γ. In this section the focus is on $\mathcal{L}_\sigma$ as a set of words, i.e. a *language*.

2.5.1 Languages and complexity

A *formal language* is a subset of the free monoid on a finite set. (See [HU], [ECHLPT] and [HRR] for background.) Our interest lies with three classes of formal languages:

$$\text{Reg} \subset \text{CF} \subset \text{Ind}$$

Reg is the class of regular languages, CF the class of context-free languages, Ind the class of indexed languages introduced by Aho [Ah]. Each of these classes is a *full abstract family of languages*, meaning that it is closed under the operations of homomorphism and inverse homomorphism between the ambient free monoids, intersections with regular languages, union, product and generation of submonoid. An important consequence of this from the point of view of group theory is that if $\mathcal{A}$ is one of these classes of languages, then the class of groups that admit a combing σ with $\operatorname{im}\sigma \in \mathcal{A}$ enjoying the fellow-traveller property (synchronous or asynchronous) is closed under the operations of free and direct products, passage to and from finite-index subgroups, and passage to quasi-convex subgroups; see [BGi1].

All three of our classes can be defined in several equivalent ways. An attractive approach is to define them in terms of generative grammars. The context-free grammars defining CF are easy to describe: there is a finite set of non-terminals $A, B, \ldots$ including the start symbol S, and a finite set of terminals (the alphabet in which the output language is written) $a, b, \ldots$; the grammar consists of a finite set of rules ("productions") by which the non-terminals generate the words of the language – the left hand side of each production consists of a single non-terminal while the right hand side consists of a word in terminals and non-terminals. When two productions have the same left hand side, it is convenient to shorten the description of the grammar by combining the two productions into one, introducing a vertical line pronounced "or" to divide the two allowable outcomes. ε is typically used to denote the empty word. For example, the context-free grammar

$$S \to P$$
$$P \to PP \mid PaP\overline{a}P \mid PbP\overline{b}P \mid P\overline{a}PaP \mid P\overline{b}PbP \mid \varepsilon$$

generates the context-free language consisting of all words in the alphabet $\{a, b, \overline{a}, \overline{b}\}$ that represent the identity in the free group $F(a, b)$ when $\overline{a}$ is interpreted as a^{-1} and $\overline{b}$ is interpreted as b^{-1}.

Indexed grammars, which arise naturally in geometry, are more complicated, allowing one to keep track of indices (see Section 2.5.6). A thorough recent treatment is given in [HRR].

The three language classes can also be defined in terms of automata (simple idealised computing machines). Each automaton takes as input words in a fixed alphabet and either *accepts* or *rejects* the word after a finite computation; the associated language is the set of accepted words. There is the following correspondence between classes of languages and the types of automata that accept them:

Reg $\leftrightarrow$ finite state automata

CF $\leftrightarrow$ pushdown automata

Ind $\leftrightarrow$ nested stack automata

In each case, one has to allow the automata to be non-deterministic; see [HU].

2.5.2 Finite state automata

A finite state automaton is a finite directed graph. The vertices are called *states*; they are divided into ACCEPT and REJECT states; there is a distinguished start state S_0. The directed edges are labelled by the alphabet A; for each state S and for each $a \in A$ there is an edge emanating from S with label a.

Given a word w in the alphabet A (imagined as being read from a one-way input tape), one traces the directed edge-path from S_0 that is labelled w. The set of words labelling paths that end at accept vertices is the accepted language of the automaton. It is a regular language, and all regular languages arise in this way. For example, the finite state automaton pictured below has accepted language $\{x^n y^m \mid n, m \in \mathbb{N}\}$.

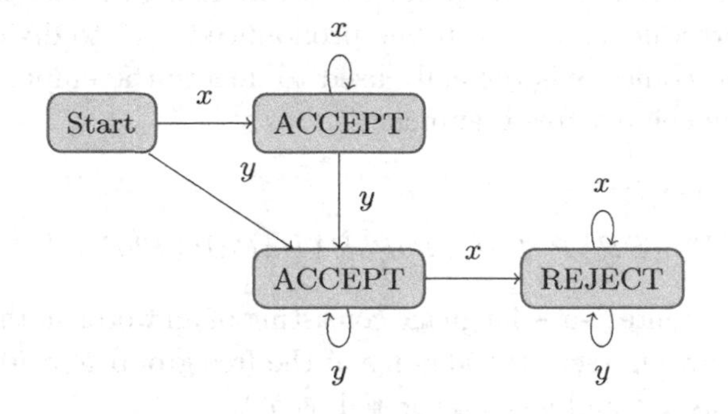

Remark 2.5.2 The reader might find it instructive to prove that given any list of finite words $u_1, \ldots, u_n$ over an alphabet A, one can construct a finite state automaton that rejects exactly those words that contain one of the u_i as a subword.

With finite state automata (FSA), if one permits non-determinism by allowing there to be more than one edge labelled a emanating from some vertices, then one does not alter the class of accepted languages. This is not true of the other language classes that we consider: in these cases one has to admit non-deterministic automata in order to obtain the full language class.

Context-free languages are the accepted languages of (non-deterministic) *pushdown automata* (PDA): these are finite state automata augmented by a single stack, so that the machine can write (using a finite stack alphabet) and erase at the top of the stack but the head cannot enter the stack. For example, by pushing a counter onto the stack each time one reads a letter a and popping one off when one reads b (changing mode with a simple FSA), one easily constructs a PDA that can recognise $\{a^n b^n \mid n \in \mathbb{N}\}$. But at the end of this calculation one has lost the stored knowledge of the exponent n, so it is not clear how to proceed to recognise $\{a^n b^n c^n \mid n \in \mathbb{N}\}$. In fact, this last language is *not* context-free. But it is an *indexed* language.

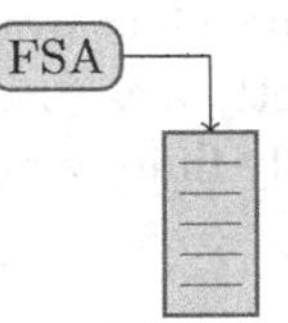

Figure 2.6 A pushdown automaton

A **nested stack automaton** is more powerful than a pushdown automaton. The head is now allowed to enter the stack in read-only mode and calculate its next move according to the stack entry it is looking at as well as the state the associated finite state automaton is in and the current letter on the input tape. The head is also able to open a new stack for calculation and to iterate this procedure, but it must close each new stack before making a move in the previous one.

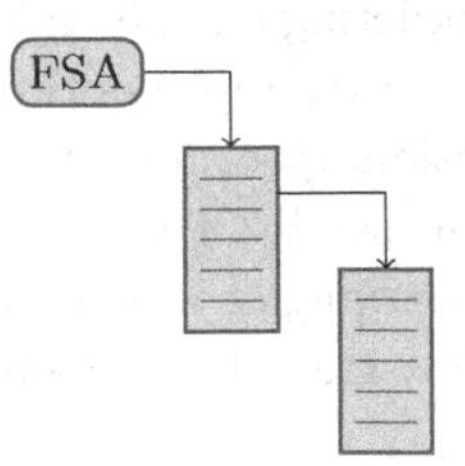

Figure 2.7 A nested stack automaton

2.5.3 Hyperbolic groups are automatic

The following is a restatement of Cannon's Theorem 2.2.13.

Theorem 2.5.3 *If Γ is hyperbolic then the language of geodesic words is regular.*

One should think of this theorem as a strong *local recognition* criterion for geodesics (*cf.* Lemma 2.2.2 and remark 2.5.2).

What is the finite state automaton M_{geod} that accepts the language of geodesics? Its states are the cone types in Γ. The **cone type** of an element γ is the set

$$C(\gamma) := \{u \in (A^{\pm})^{*} \mid d(1, \gamma u) = d(1, \gamma) + |u|\}.$$

More intuitively, $C(\gamma)$ is the view from the vertex γ in the Cayley graph of Γ as one looks directly away from the identity. The important observation that there are only finitely many views is due to Cannon [C] (see [BHa1, p.455]).

Proposition 2.5.4 *With respect to any finite generating set, a hyperbolic group has only finitely many cone types.*

Consider the finite state automaton whose states are the cone types for a hyperbolic group Γ with a fixed finite generating set A together with a single fail state, with an edge joining $C(\gamma)$ to $C(\gamma a)$ for each $a \in C(\gamma)$ and an edge to the fail state if $a \notin C(\gamma)$, and with initial state $C(1)$. The regular language accepted by this automaton is the set of geodesic words in Γ, so Theorem 2.5.3 is proved.

From the language of all geodesics, one can extract a regular sublanguage with a unique geodesic representing each element of the group by simply imposing a linear ordering on the alphabet $A^{\pm}$ and then selecting, for each $\gamma \in \Gamma$, the geodesic σ_γ representing γ that comes before any other in the resulting dictionary order. (It is not hard to see that this sublanguage is still regular; see [ECHLPT, p.57].)

The thinness of geodesic triangles in hyperbolic groups assures us that σ satisfies the 2-sided fellow-traveller property, so we have

Corollary 2.5.5 *Hyperbolic groups are biautomatic.*

2.5.4 Local recognition of geodesics

We commented above that Cannon's Theorem 2.2.13 should be regarded as a strong local criterion for recognising geodesics. Focussing on this idea (and extending Remark 2.5.2) is easy to see that the geodesic words in an arbitrary basis of a free abelian group also form a regular language. A more sophisticated version of the same argument, based on an understanding of combinatorial geodesics in CAT(0) cube complexes [NiR] and systolic complexes [Sw], allowed Niblo–Reeves and Swiatkowski to prove far reaching generalisations for groups acting freely and cocompactly by isometries on these complexes, from which it can be deduced that these groups are biautomatic. Swiatkowski's work, in particular, brings into stark and explicit focus the connection between local recognition of global geodesics (or other preferred families of paths in the geometry) and automaticity.

2.5.5 Effective computation

The development of software to draw Cayley graphs of hyperbolic groups is intimately connected with the development of the theory of automatic groups. Indeed a key feature of automatic groups is that their amenability to computation is by no means just theoretical: Derek Holt, in particular, has developed many effective computational tools that enable one to find and work with automatic structures in many contexts; see [Ho].

2.5.6 Indexed languages in geometry

Indexed languages arose naturally in my work with Gilman on the complexity of normal forms for 3-manifold groups [BGi1]. Their utility in this context is closely linked to the following much-studied language of words (which enjoys many rich interpretations).

We define the language $\mathcal{L}_{\mathbb{E}}$ of *crossing sequences* to be the set of words obtained by drawing straight lines of rational slope emanating from the origin in the Euclidean plane and recording the order in which they cross horizontal and vertical lines with integer coordinates. (One perturbs lines up slightly to avoid ambiguity for lines ending at vertices (m, n) with m and n not coprime.) For example, the line below has crossing sequence $xyxyxyxxyxyx$.

Theorem 2.5.6 *$\mathcal{L}_{\mathbb{E}}$ is an indexed language.*

The reader may find it instructive to consider why $\mathcal{L}_{\mathbb{E}}$, the language of crossing sequences, is not context-free. In fact, a context-free language can only ever describe finitely many slopes in the plane (cf. [BGi1, Proposition 1.3]).

Indexed languages and reparametrisation. The known constructions of combable groups that are not automatic or else exhibit extreme

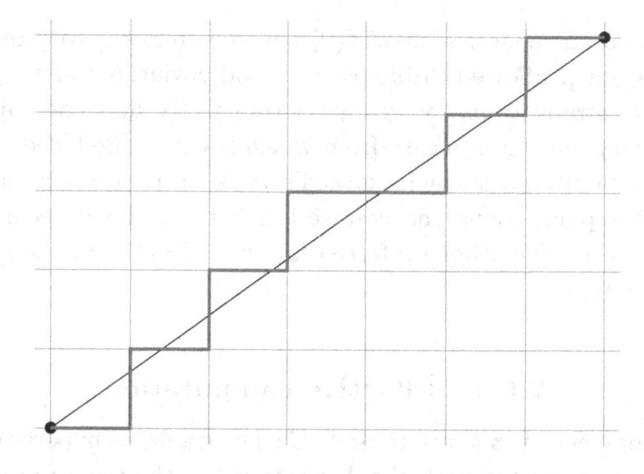

behaviour (e.g. Examples 2.4.9, 2.4.10 and [Bri7], [Bri6]) tend to involve taking normal forms that enjoy the asynchronous fellow-traveller property and then adapting them with suitable global reparametrisations. Indexed languages are well-adapted to such reparametrisation processes, whereas regular and context-free languages tend to be too inflexible, with the *pumping lemma* (as described in [HU] or [HRR]) providing a key obstruction.

2.6 Examples

The following construction provides a rich source of semihyperbolic groups that are not CAT(0); we will give a specific example from low-dimensional topology.

2.6.1 Central extensions of hyperbolic groups

For any group Q, extensions $1 \to A \to E \to Q \to 1$ with A central are classified by classes in $H^2(Q, A)$. If Q is hyperbolic and A is finitely generated, then every class in $H^2(Q, A)$ can be represented by a cocycle $\sigma : Q \times Q \to A$ whose image is finite. In this case, it is easy to see that E will be quasi-isometric to $A \times Q$, but they will not be commensurable in general (see Example 2.6.2).

Now, $A \times Q$ is combable by quasi-geodesics (since A and Q are) and the geometry of a (non-equivariant) combing is preserved when the combing is transported by a quasi-isometry, so E also has such a combing. It is less obvious that it is semihyperbolic, but this can be proved by analysing the obvious bicombing directly; see [AB]. More delicately, Neumann and Reeves [NeR] proved that the cocycle defining E can be described by a finite state automaton, and thus they proved:

Theorem 2.6.1 *If Q is hyperbolic, A is finitely generated and $1 \to A \to E \to Q \to 1$ is a central extension, then E is biautomatic.*

The group E will in general not be CAT(0) because A will not be a direct factor of a subgroup of finite index in E, whereas Theorem 2.3.1 tells us that it would be if E were to act properly by semisimple isometries on a CAT(0) space.

Distinguishing Example 2.6.2 Let Σ_g be a closed hyperbolic surface of genus at least 2. Let M_g^3 be the total space of the unit tangent bundle of Σ_g; this is a closed 3-manifold foliated by circles in an obvious manner. The free homotopy class of the circle fibres is central in $\pi_1 M_g^3$ and there is a short exact sequence

$$1 \to \mathbb{Z} \to \pi_1 M_g^3 \to \pi_1 \Sigma_g \to 1.$$

This short exact sequence is a central extension and does not split, even on passing to subgroups of finite index, since $\pi_1 M_g^3$ is a cocompact lattice in the Lie group $\widetilde{\mathrm{PSL}(2,\mathbb{R})}$, which does not contain the fundamental group of any closed surface of genus at least 2. It follows that whenever $\pi_1 M_g^3$ acts by semisimple isometries on a CAT(0) space, its centre $Z \cong \mathbb{Z}$ must fix a point, for otherwise the generator ζ of Z would be a hyperbolic isometry and Theorem 2.3.1(4) would imply that the above sequence virtually split.

Thus $\pi_1 M_g^3$ provides us with a natural example of a group that is semihyperbolic (by Theorem 2.6.1) but is not CAT(0).

2.6.2 Placing the gems

A major achievement of the foundational work on automatic groups by Epstein *et al.* [ECHLPT] was to determine which **3-manifold groups** are automatic. Subsequently, Bridson [Bri2] showed that all 3-manifold groups are asynchronously combable, and determining the linguistic complexity of those combings was the primary motivation in Bridson and Gilman's exploration of formal language theory in the context of geometry [BGi1]. The following is a compilation of these results. We refer the reader to [AFW] for a comprehensive account of the structure of 3-manifold groups.

Theorem 2.6.3 *1 The fundamental group of every compact 3-manifold M is asynchronously combable, with an* indexed *combing.*

2 Lattices in the 3-dimensional Lie group Nil *are not asynchronously combable with a* context-free *combing.*

3 If M does not have a connected summand that is finitely covered by a torus bundle over the circle with infinite holonomy, then $\pi_1 M$ is automatic and semihyperbolic;

4 condition (3) is equivalent to requiring that $\pi_1 M$ does not have a subgroup of the form $\mathbb{Z}^2 \rtimes_\psi \mathbb{Z}$ with ψ infinite, and it is also equivalent to requiring that the Dehn function of $\pi_1 M$ is linear or quadratic.

Example 2.6.2 shows that not all semihyperbolic 3-manifold groups are CAT(0). In order to determine which are, one has to examine the JSJ decomposition of the manifold: Kapovich and Leeb [KL] (after [Le]) gave the definitive answer for manifolds that are closed or have toral boundary, and Bridson extended this to the general case in [Bri11]. A deeper result of Pryztycki and Wise [PW], relying on Agol's Theorem [Ag], shows that a compact aspherical 3-manifold M admits a metric of non-positive curvature if and only if $\pi_1 M$ is virtually special in the sense of Haglund and Wise [HaglW].

Mapping class groups and $\mathrm{Out}(F_n)$**.** Despite enjoying various algebraic properties associated to non-positive curvature, outer automorphism groups of free groups do not fit happily into any assessment of groups that is based on convexity properties of combings – see [BV1]. And their Dehn functions are exponential [HM], [BV2]. However, these groups are acylindrically hyperbolic (see [Os2] and references therein). Thus $\mathrm{Out}(F_n)$ provides a natural and striking instance of the fact that different approaches to exploring manifestations of non-positive curvature in group theory capture markedly different classes of groups.

In contrast to $\mathrm{Out}(F_n)$, mapping class groups of hyperbolic surfaces fit well into both the semihyperbolic and acylindrically hyperbolic frameworks. Let $\mathrm{Map}(\Sigma_g)$ denote the mapping class group of a closed surface of genus g, i.e. the group of isotopy classes of orientation-preserving homeomorphisms of Σ_g. Mosher [M] proved that $\mathrm{Map}(\Sigma_g)$ is automatic and Hamenstädt [Ham1] described a biautomatic structure.

For $g \geq 3$, $\mathrm{Map}(\Sigma_g)$ is another naturally occurring example of a semihyperbolic group that is not CAT(0). Indeed $\mathrm{Map}(\Sigma_g)$ contains the group $\pi_1 M^3_{g-1}$ considered in Example 2.6.2 and therefore cannot even act properly by semisimple isometries on a CAT(0) space if $g \geq 3$; see [KL] and [BHa1, p.257].

To see that $\mathrm{Map}(\Sigma_g)$ contains $\pi_1 M^3_{g-1}$, consider a simple loop c on the surface Σ_g that separates Σ_g into two components, one of genus $g-1$; let S be this component. The subgroup $\Gamma < \pi_1 M^3_{g-1}$ consisting of isotopy classes with a representative supported on S has cyclic centre Z generated by the Dehn twist ζ in c. The quotient of Γ by Z is the mapping class group of a punctured surface of genus $g-1$, which contains a copy of $\pi\Sigma_{g-1}$, namely the homeomorphisms that push the puncture around loops on the surface (see, for example, [FM, p.149]). The inverse image of this surface group in Γ is the isomorphic copy of $\pi_1 M^3_{g-1}$ that we were seeking.

More subtle restrictions on the actions of $\mathrm{Out}(F_n)$ and $\mathrm{Map}(\Sigma_g)$ on CAT(0) spaces can be found in [Bri4], [Bri5].

Lattices in semisimple Lie groups. Non-uniform lattices in semisimple Lie groups other than $\mathrm{SO}(n,1) = \mathrm{Isom}^+(H^n)$ contain polycyclic subgroups that are not virtually abelian, and hence by Theorem 2.4.5 cannot be semihyperbolic. One can pursue the question of what other sorts of combings they admit, but a far more natural thing to do in the rank 1 case is to consider such groups in the context of relative hyperbolicity.

Non-uniform lattices $\Gamma < \mathrm{SO}(n,1)$ act properly and cocompactly on a natural CAT(0) space, namely the result of removing an equivariant system of disjoint horoballs about the parabolic fixed points of Γ in compactified hyperbolic space $\mathbb{H}^n \sqcup \partial\mathbb{H}^n$. A more subtle argument, due to Epstein *et al.* [ECHLPT] shows that Γ is also automatic.

Uniform lattices are obviously CAT(0), since they act properly and cocompactly by isometries on the symmetric space of the associated semisimple Lie group. There seems little reason to believe that irreducible lattices in higher rank Lie groups must be automatic, but this question remains open.

The task of determining the Dehn functions of non-uniform lattices is a rich and delicate one that has received much attention: examples including $\mathrm{SL}(n,\mathbb{Z})$ contribute to the zoo of groups with quadratic Dehn functions [Y2].

Free-by-cyclic groups. Free-by-cyclic groups $\Gamma(\phi) = F_n \rtimes_\phi \mathbb{Z}$ provide a rich testing ground in the context of semihyperbolicity and its variations. If no power of $\phi \in \mathrm{Out}(F_n)$ leaves a conjugacy class in F_n invariant, then $\Gamma(\phi)$ is hyperbolic [Brin], [BF1]. Hagen and Wise recently proved that in this case $\Gamma(\phi)$ can also be cubulated, and hence is CAT(0) and virtually-special [HageW]. On the other hand, Gersten [Ge5] proved that for certain automorphisms ϕ of polynomial growth, $\Gamma(\phi)$ is not CAT(0). In his influential problem list [Ge3], Gersten asked if $\Gamma(\phi)$ was always automatic, but this was later proved not to be the case: [BrB2], [BR2]. As yet, there is no definitive classification as to which $\Gamma(\phi)$ are CAT(0) or automatic. The non-hyperbolic examples all have quadratic Dehn functions [BGr].

2.7 Algorithmic construction of classifying spaces

We saw in Section 2.2.4 that torsion-free hyperbolic groups have finite classifying spaces. Indeed the construction that we described there gives an algorithm that, given a finite presentation of a torsion-free δ-hyperbolic group Γ, with knowledge of the constant δ, will construct a finite $K(\Gamma, 1)$, namely the quotient of the Rips complex by Γ. Force is lent to this by the fact there is a partial algorithm that, given a finite presentation of a hyperbolic group will calculate δ; see [P1]. Torsion-free CAT(0) groups also have finite classifying spaces, but there is no obvious algorithm that, given a finite presentation of the group, will construct such a space. For more general semihyperbolic and combable groups, one knows only that the groups are of type FP_∞.

Given this context, a reasonable hope would be that there is an algorithm that, given a finite presentation of a combable group Γ, will construct any specified skeleton of a $K(\Gamma, 1)$. Comparing with the hyperbolic situation, one expects that the fellow-traveller constant will be required in the construction but hopes that in favourable situations the constant might be calculated directly from the presentation by a partial algorithm. The following theorem, the main result of [BR1], realises this hope. The distinction between automatic and combable groups in this theorem lies with the fact that there is an algorithm that, given a finite presentation of an automatic group, will calculate an explicit automatic structure [ECHLPT] from which one can then calculate an explicit fellow-traveller constant [ECHLPT, Lemma 2.3.2]. The following is a weak form of Theorem 3.3 of [BR1].

Theorem 2.7.1 *There is an algorithm that given a finite presentation* *of an automatic group and an integer d, will construct the d-skeleton of a classifying space $K(G,1)$ for the group presented.*

There is also an algorithm that will achieve the same construction when given a finite presentation of a combable group together with an explicit fellow-traveller constant.

Proceeding with care, one can use an algorithm that constructs classifying spaces to calculate the (co)homology of the groups concerned.

Corollary 2.7.2 *There is an algorithm that, given a finite presentation of an automatic group G, a finitely generated abelian group A, and an integer d, will calculate $H_d(G,A)$ and $H^d(G,A)$.*

If one reflects on the proof that combable groups are finitely presented and the proof that the Rips complex for a hyperbolic group is contractible, one quickly gets an idea of how a proof of the above theorem might proceed – roughly, one expects merely to add all i-cells up to dimension d whose vertex set in the Cayley graph is of bounded diameter, where the bound depends on the fellow-traveller constant. However, this naive approach quickly runs into difficulties and the actual construction is considerably more subtle (although it proceeds along morally the same lines).

2.8 Cubulated groups and systolic groups

CAT(0) cube complexes played a central role in the development of the theory of CAT(0) geometry, providing a method for constructing high-dimensional examples (with a combinatorial criterion for verifying non-positive curvature – the *link condition*). They served both as a testing ground for the general theory and as a context for constructions that settled old problems in geometry and topology – including, notably, Mike Davis's construction of closed aspherical manifolds that are not covered by Euclidean space [Da] (see [BHa1, p.213]).

The transformational work of Dani Wise and his coauthors on cubulation ([HaglW], [W1] *et seq.*) has added many new layers to the theory, endowing it with large colonies of examples and, again, solving fundamental problems in geometry and topology. The widely-applicable techniques that have been developed for cubulating groups bring a rich supply of new examples to the pantheon of semihyperbolic groups. Moreover, all of these groups are biautomatic [NiR].

The complementary theory of systolic complexes and groups, starting from [JS] (also [Hag]) and developed largely by the Polish school, provides a further rich source of semihyperbolic groups. Systolic complexes are defined by a combinatorial condition on links that ensures behaviour that is reminiscent of CAT(0) geometry but quite distinct from it. In general, one does not expect groups that act properly and cocompactly on systolic complexes to be CAT(0), but they are biautomatic [Sw].

2.9 Subgroups

Many of the ideas explained in these lectures originated in the late 1980s or early 1990s, but at that time little was known about the subgroup structure of semihyperbolic groups (cf. [Bri11]), although there were clear indications of its richness, such as the work of Bestvina and Brady [BeB]. Since then, following intensive work by many authors, far more structure has emerged: in some settings, notably residually free groups [BHMS], this structure is subtle and diverse but largely understood, while in other settings, such as direct products of arbitrary hyperbolic groups, all manner of wildness has been uncovered – for example one can construct hyperbolic groups Γ that are CAT(0), but for which there is no algorithm to decide isomorphism among the finitely presented subgroups of $\Gamma \times \Gamma \times \Gamma$ – see [BM], [Bri1] and references therein. The subsequent explosion of work on cubulated groups provides many further examples, and it also provides technology that allows one to demonstrate that the finitely presented subgroups of classical groups such as mapping class groups is far more complex than previously imagined – see [Bri9].

In the more restricted setting of hyperbolic groups, we still know remarkably little about the nature of finitely presented subgroups. The celebrated construction of Rips [R] allows one to manufacture a wide array of finitely generated subgroups that are not finitely presented, but we know of very few finitely presented subgroups that are not hyperbolic. In cohomological dimension 2 there are no such subgroups [Ge4]. And for many years Noel Brady's example [Bra] was the only such subgroup known. More recently, a novel construction by Yash Lodha [Lo] and the more sweeping constructions by Robert Kropholler in his thesis [K] have helped correct this situation. Kropholler's work, in particular, gives hope that we shall soon have examples of subgroups of hyperbolic groups $H < G$ such that H has a finite classifying space $K(H,1)$ but is not hyperbolic, but for the moment no such groups have been found.

2.10 Containments

The various classes of groups that we have discussed are arranged in the following diagram. Recent results of Leary and Minasyan [LM] provide the first examples of CAT(0) groups that are not biautomatic. For the moment, we have no example of an automatic group that is not biautomatic. Here we abbreviate by *Combable q.g.* the property of being combable with quasi-geodesics as combing lines.

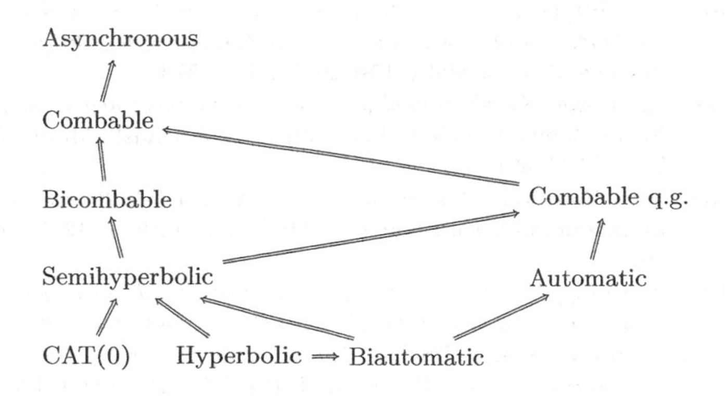

References

[Ag] I. Agol, *The virtual Haken conjecture* (with appendix by I. Agol, D. Groves and J. Manning), Documenta Math. **18**(2013), 1045–1087.

[Ah] A.V. Aho, *Indexed grammars – an extension of context-free grammars*, Journal of the ACM, **15** (1968), 647–671.

[Al] J. Alonso, *Combings of groups*, in "Algorithms and classification in combinatorial group theory" (Berkeley, CA, 1989), pp. 165–178, Math. Sci. Res. Inst. Publ. **23**, Springer, New York, 1992.

[AB] J. Alonso and M.R. Bridson, *Semihyperbolic groups*, Proc. London Math. Soc., **70** (1995), 56–114.

[AFW] M. Aschbrenner, S. Friedl, H. Wilton, *3-Manifold Groups*, EMS Series of Lectures in Mathematics, European Math. Soc., 2015.

[BBMS1] G. Baumslag, M.R. Bridson, C.F. Miller III, H. Short, *Subgroups of automatic groups and their isoperimetric functions*, J. London Math. Soc., **56**(1997), 292–304.

[BBMS2] G. Baumslag, M.R. Bridson, C.F. Miller III, H. Short, *Fibre products, non-positive curvature, and decision problems*, Comment. Math. Helv. **75**(2000), 457–477.

[BGSS] G. Baumslag, S.M. Gersten, M. Shapiro, H. Short, *Automatic groups and amalgams*, J. Pure Appl. Algebra **76**(1991), 229–316.

[BeB] M. Bestvina and N. Brady, *Morse theory and finiteness properties of groups*, Invent. Math. **129**(1997), 445–470.

[BBF] M. Bestvina, K. Bromberg, K. Fujiwara, *Constructing group actions on quasi-trees and applications to mapping class groups*, Publ. Math. IHÉS **122**(2015), 1–64.

[BF1] M. Bestvina and M. Feighn, *A combination theorem for negatively curved groups*, J. Differential Geom. **35**(1992), 85–101.

[BF2] M. Bestvina and M. Feighn, *A hyperbolic* $\mathrm{Out}(F_n)$*-complex*, Groups Geom. Dyn. **4** (2010), 31–58.

[BORS] J-C. Birget, A. Yu. Ol'shanskii, E. Rips, M.V. Sapir, *Isoperimetric functions of groups and computational complexity of the word problem*, Ann. of Math. **156**(2002), 467– 518.

[Bor] V. Borisov, *Simple examples of groups with unsolvable word problem*, Math. Zametki **6**(1969), 521–532; English transl., Math. Notes 6, 768–775 (1969).

[Bow] B.H. Bowditch, *A short proof that a subquadratic isoperimetric inequality implies a linear one*, Michigan Math. J. **42**(1995), 103–107.

[Bra] N. Brady, *Branched coverings of cubical complexes and subgroups of hyperbolic groups*, J. London Math. Soc. **60**(1999), 461–480.

[BrB1] N. Brady and M.R. Bridson, *There is only one gap in the isoperimetric spectrum*, Geom. Funct. Anal. (GAFA), **10** (2000), 1053-1070.

[BrB2] N. Brady and M.R. Bridson, *On the absence of biautomaticity in certain graphs of abelian groups*, preprint 1998.

[BBFS] N. Brady, M.R. Bridson, M. Forester, K. Shankar, *Snowflake groups, Perron-Frobenius eigenvalues and isoperimetric spectra*, Geom. Topol. **13**(2009), 141–187.

[Bri1] M.R. Bridson, *Non-positive curvature and complexity for finitely presented groups*, International Congress of Mathematicians. Vol. II, 961–987, Eur. Math. Soc., Zurich 2006.

[Bri2] M.R. Bridson, *Combing semidirect products and 3-manifold groups*, Geom. Funct. Anal. (GAFA), **3** (1993), 263–278.

[Bri3] M.R. Bridson, *On the geometry of normal forms in discrete groups*, Proc. London Math. Soc., **67** (1993), 516–616.

[Bri4] M.R. Bridson, *On the dimension of CAT(0) spaces where mapping class groups act*, J. Reine Angew. Math. **673** (2012), 55–68.

[Bri5] M.R. Bridson, *The rhombic dodecahedron and semisimple actions of* $Aut(F_n)$ *on CAT(0) spaces*, Fund. Math. **214** (2011), 13–25.

[Bri6] M.R. Bridson, *The conjugacy and isomorphism problems for combable groups*, Math. Ann., **327** (2003), 305–314.

[Bri7] M.R. Bridson, *Combings of groups and the grammar of reparameterisation*, Comment. Math. Helv., **78** (2003), 752-771.

[Bri8] M.R. Bridson, *On the growth of groups and automorphisms*, Intl. J. Alg. Comp. **15** (2005), 869–874.

[Bri9] M.R. Bridson, *On the subgroups of right-angled Artin groups and mapping class groups*, Math. Res. Lett., **20** (2013), 203–212.

[Bri10] M.R. Bridson, *The geometry of the word problem*, in "Invitations to geometry and topology", (M.R. Bridson and S.M. Salamon, eds.), OUP, Oxford 2002, pp. 29-91.

[Bri11] M.R. Bridson, *On the subgroups of semihyperbolic groups*, in "Essays on geometry and related topics", pp. 85–111, Monogr. Enseign. Math. **38**, Geneva, 2001.

[Bri12] M.R. Bridson, *Polynomial Dehn functions and the length of asynchronous automatic structures*, Proc. London Math. Soc. 85 (2002), 441–466.

[BGi1] M.R. Bridson and R. Gilman, *Formal language theory and the geometry of 3-manifolds*, Comm. Math. Helv., **71** (1996), 525–555.

[BGi2] M.R. Bridson and R. Gilman, *Context-free languages of subexponential growth*, J. Comput. System Sci. 64 (2002), 308–310.

[BGr] M.R. Bridson and D. Groves, *The quadratic isoperimetric inequality for mapping tori of free group automorphisms*, Mem. Amer. Math. Soc. **203**(2010), no. 955.

[BHa1] M.R. Bridson and A. Haefliger, Metric Spaces of Non-Positive Curvature, Grundlehren der Mathematischen Wissenschaften, Vol. 319, Springer-Verlag, Heidelberg-Berlin, 1999.

[BHa2] M.R. Bridson and J. Howie, *Conjugacy of finite subsets in hyperbolic groups*, Internat. J. Algebra Comput. 15(4) (2005) 725–756.

[BHMS] M.R. Bridson, J. Howie, C.F. Miller III, and H. Short, *On the finite presentation of subdirect products and the nature of residually free groups*, (with J. Howie, C.F. Miller III, and H. Short), American J. Math., **135** (2013), 891–933.

[BM] M.R. Bridson and C.F. Miller III, *Recognition of subgroups of direct products of hyperbolic groups*, Proc. Amer. Math. Soc. **132** (2004), 59–65.

[BR1] M.R. Bridson and L. Reeves, *On the algorithmic construction of classifying spaces and the isomorphism problem for biautomatic groups*, Science China Mathematics (volume for Fabrizio Catanese) **54** (2011), 1533–1545.

[BR2] M.R. Bridson and L. Reeves, *On the absence of automaticity in certain free-by-cyclic groups*, preprint 2006.

[BV1] M.R. Bridson and K. Vogtmann, *On the geometry of the automorphism group of a free group*, Bull. London Math. Soc. **27**(1995) 544–552.

[BV2] M.R. Bridson and K. Vogtmann, *The Dehn functions of* $\mathrm{Out}(F_n)$ *and* $\mathrm{Aut}(F_n)$, Ann. Inst. Fourier, **62** (2012), 1811–1817.

[Brin] P. Brinkmann, *Hyperbolic automorphisms of free groups*, Geom. Funct. Anal. **10**(2000), 1071–1089.

[BH] D.J. Buckley and D.F. Holt, *The conjugacy problem in hyperbolic groups for finite lists of group elements*, Internat. J. Algebra Comput. **35**(2013), 1127–1150.

[C] J.W. Cannon, *The combinatorial structure of cocompact discrete hyperbolic groups*, Geom. Dedicata **16** (1984), 123148.

[CM] D.J. Collins, and C.F. Miller III., *The conjugacy problem and subgroups of finite index*, Proc. London Math. Soc. **34**(1977), 535–556.

[DGr] F. Dahmani and D. Groves, *The isomorphism problem for toral relatively hyperbolic groups*, Publ. Math. IHÉS **107**(2008), 211–290.

[DGu] F. Dahmani and V. Guirardel, *The isomorphism problem for all hyperbolic groups*, Geom. Funct. Anal. **21**(2011), 223–300.

[Da] M.W. Davis, *Groups generated by reflections and aspherical manifolds not covered by Euclidean space*, Ann. of Math. **117**(1983), 293–324.

[Deh] M. Dehn, *Papers on Group Theory and Topology*, translated and introduced by John Stillwell, Springer Verlag, Berlin, Heidelberg, New York, 1987.

[Del] T. Delzant, *Sous-groupes distingueés et quotients des groupes hyperboliques*, Duke Math. J. **83** (1996), 661–682.

[DERY] W. Dison, M. Elder, T.R. Riley, R. Young, *The Dehn function of Stallings' group*, Geom. Funct. Anal. **19**(2009), 406–422.

[ECHLPT] D.B.A. Epstein, J.W. Cannon, D.F. Holt, S.V.F. Levy, M.S. Paterson, W.P. Thurston, Word processing in groups. Jones and Bartlett Publishers, Boston, MA, 1992.

[EH] D.B.A. Epstein and D.F. Holt, *The linearity of the conjugacy problem in word-hyperbolic groups*, Internat. J. Algebra Comput. **16**(2006), 287–305.

[FM] B. Farb and D. Margalit, *A Primer on Mapping Class Groups*, Princeton Univ. Press, Princeton NJ, 2012.

[FJ1] F.T. Farrell and L.E. Jones, *A topological analogue of Mostow's rigidity theorem*, J. Amer. Math. Soc. **2**(1989), 257–370.

[FJ2] F.T. Farrell and L.E. Jones, *Rigidity and other topological aspects of compact nonpositively curved manifolds*, Bull. Amer. Math. Soc. **22**(1990), 59–64.

[FO] T. Fukaya and S-I. OguniI, *A coarse Cartan-Hadamard theorem with applications to the coarse Baum-Connes conjecture*, ARXIV:1705.05588.

[Ge1] S.M. Gersten, *Bounded cohomology and combings of groups*, `http://www.math.utah.edu/~sg/Papers/bdd.pdf`, Preprint 1991.

[Ge2] S.M. Gersten, *Finiteness properties of asynchronously automatic groups*, in "Geometric Group Theory", Ohio State University Mathematical Research Institute Publications, R. Charney, M. Davis, M. Shapiro (eds.), de Gruyter, New York, 1995.

[Ge3] S.M. Gersten, *Problems on automatic groups*, in "Algorithms and classification in combinatorial group theory" (Berkeley, CA, 1989), pp. 225–232, Math. Sci. Res. Inst. Publ. vol. 23, Springer, New York, 1992.

[Ge4] S.M. Gersten, *Subgroups of word hyperbolic groups in dimension 2*, J. London Math. Soc. **54**(1996), 261–283.

[Ge5] S.M. Gersten, *The automorphism group of a free group is not a* CAT(0) *group*, Proc. Amer. Math. Soc. **121**(1994), 999–1002.

[GS] S.M. Gersten and H.B. Short, *Rational subgroups of biautomatic groups*, Ann. of Math. **134**(1991), 125–158.

[Gr1] M. Gromov, *Hyperbolic groups*, in "Essays in Group Theory", pp. 75–263, Math. Sci. Res. Inst. Publ. vol. 8, Springer, New York, 1987.

[Gr2] M. Gromov, *Asymptotic invariants of infinite groups*, in "Geometric group theory, Vol. 2 (Sussex, 1991)", pp. 1295, London Math. Soc. Lecture Note Ser., 182, Cambridge Univ. Press, Cambridge, 1993.

[Gu] V.S. Guba, *The Dehn function of Richard Thompson's group F is quadratic*, Invent. Math. **163**(2006), 313–342.

[HageW] M.F. Hagen and D.T. Wise, *Cubulating hyperbolic free-by-cyclic groups: the general case*, Geom. Funct. Anal. **25**(2015), 134–179.

[Hag] F. Haglund, *Complexes simpliciaux hyperboliques de grande dimension*, preprint, Orsay 2003.

[HaglW] F. Haglund and D.T. Wise, *Special cube complexes*, Geom. Funct. Anal. **17**(2008), 1551–1620.

[Ham1] U. Hamenstädt, *Actions of the mappings class group*, Proceedings of the International Congress of Mathematicians. Vol. II, pp. 1002–1021, Hindustan Book Agency, New Delhi, 2010.

[Ham2] U. Hamenstädt, *Geometry of the mapping class group II: A biautomatic structure*, ARXIV:0912.0137.

[HM] M. Handel and L. Mosher, *Lipschitz retraction and distortion for subgroups of* $\mathrm{Out}(F_n)$, Geom. Topol. **17**(2013), 1535–1579.

[HU] J.E. Hopcroft and J.D. Ullman, An Introduction to Automata Theory, Languages, and Computation, Addison-Wesley, Boston MA, 1979.

[Ho] D.J. Holt, *The Warwick automatic groups software*, in Geometrical and Computational Perspectives on Infinite Groups, DIMACS Series in Discrete Mathematics and Theoretical Computer Science, vol. 25, ed. Gilbert Baumslag et. al., pp. 69–82.

[HRR] D.J. Holt, S. Rees, C.E. Röver, Groups, languages and automata, London Mathematical Society Student Texts, 88, Cambridge University Press, Cambridge, 2017.

[JS] T. Januszkiewicz, J. Świątkowski, *Simplicial nonpositive curvature*, Publ. Math. IHES, **104**(2006), 1–85.

[KL] M. Kapovich and B. Leeb, *Actions of discrete groups on nonpositively curved spaces*, Math. Ann. **306**(1996), 341–352.

[K] R. Kropholler, *Finiteness properties and* CAT(0) *groups*, DPhil Thesis, University of Oxford, 2016,

[LM] I.J. Leary and A. Minasyan, in preparation.

[Le] B. Leeb, *3-manifolds with(out) metrics of nonpositive curvature*, Invent. Math. **122**(1995), 277–289.

[Lo] Y. Lodha, *A hyperbolic group with a finitely presented subgroup that is not of type* FP_3, ARXIV:1403.6716V2

[MS] D. Meintrup and T. Schick, *A model for the universal space for proper actions of a hyperbolic group*, New York J. Math. **8** (2002), 1–7.

[M] L. Mosher, *Mapping class groups are automatic*, Ann. of Math. **142**(1995), 303–384.

[NeR] W.D. Neumann and L. Reeves, *Central extensions of word hyperbolic groups*, Annals of Math. **145**(1997), 183–192.

[NiR] G.A. Niblo and L. Reeves, *Groups acting on* CAT(0) *cube complexes*, Geom. Topol. **1**(1997), 1–7.

[Ol1] A. Yu. Ol'shanskii, *Almost every group is hyperbolic*, Internat. J. Algebra Comput. **2**(1992), 1–17.

[Ol2] A. Yu. Ol'shanskii, *Hyperbolicity of groups with subquadratic isoperimetric inequality*, Internat. J. Algebra Comput. **1**(1991), 281–289.

[Ol3] A. Yu. Ol'shanskii, *Polynomially-bounded Dehn functions of groups*, ARXIV:1710.00550.

[OlS] A. Yu. Ol'shanskii and M.V. Sapir, *Conjugacy problem in groups with quadratic Dehn function*, ARXIV:1809.00280.

[Os1] D. Osin, *Acylindrically hyperbolic groups*, Trans. Amer. Math. Soc. **368** (2016), 851–888.

[Os2] D. Osin, *Groups acting acylindrically on hyperbolic spaces*, ARXIV:1712.00814, to appear in Proc. ICM Rio de Janeiro 2018.

[P1] P. Papasoglu, *An algorithm detecting hyperbolicity*, Geometric and computational perspectives on infinite groups (Minneapolis, MN and New Brunswick, NJ, 1994), pp. 193–200, DIMACS Ser. Discrete Math. Theoret. Comput. Sci., 25, 1996.

[P2] P. Papasoglu, *On the sub-quadratic isoperimetric inequality*, Geometric group theory (Columbus, OH, 1992), 149–157, Ohio State Univ. Math. Res. Inst. Publ., 3, de Gruyter, Berlin, 1995.

[PW] P. Przytycki and D.T. Wise, *Mixed 3-manifolds are virtually special*, J. Amer. Math. Soc. **31**(2018), 319–347.

[R] E. Rips, *Subgroups of small cancellation groups*, Bull. London Math. Soc. **14**(1982), 45–47.

[Se] Z. Sela, *The isomorphism problem for hyperbolic groups I*, Ann. Math. **141**(1995), 217–283.

[Sh] H. Short, *Groups and combings*, ENS Lyon, 1989.

[Sw] J. Swiatkowski, *Regular path systems and (bi)automatic groups*, Geometriae Dedicata **118**(2006), 23–48.

[W1] D.T. Wise, *The Structure of Groups with a Quasiconvex Hierarchy*, preprint, McGill 2011.

[W2] D.T. Wise, *From Riches to Raags: 3-manifolds, Right-angled Artin Groups, and Cubical Geometry*, CBMS Lectures notes, 134pp, to appear.

[Y1] R. Young, *Filling inequalities for nilpotent groups through approximations*, Groups Geom. Dyn. **7**(2013), 977–1011.

[Y2] R. Young, *The Dehn function of* $\mathrm{SL}(n,\mathbb{Z})$, Ann. of Math. **177**(2013), 969–1027.

3

Acylindrically hyperbolic groups

Benjamin Barrett
School of Mathematics
University of Bristol
University Walk
Bristol BS8 1TW
United Kingdom

Abstract

These are notes from a mini-course lectured by Denis Osin on acylindrically hyperbolic groups given at *Beyond Hyperbolicity* in Cambridge in June 2016.

In this course we will be looking at acylindrically hyperbolic groups. Such groups make up a large portion of the "universe of finitely presented groups" described in Martin Bridson's course. Through examples of such groups we shall see that this class is indeed a large portion of the universe, but we shall also see that there is much we can prove. In particular, we will discuss results on small-cancellation theory and Dehn fillings, as well as some newer results.

3.1 Acylindrically hyperbolic groups

We begin with some definitions and basic results. Fix a hyperbolic metric space (X, d_X) with base point x. We do not assume that X is proper. Let ∂X be the sequential boundary of X with respect to the base point.

Fix also a group G that acts on X by isometries; this action extends to an action on $X \cup \partial X$. There is a classification of elements of G in terms of their action on X:

1 $g \in G$ is *elliptic* if $\langle g \rangle$ has bounded orbits.

2 g is *loxodromic* if $n \to g^n x$ is a quasi-isometric embedding $\mathbb{Z} \to X$. Equivalently, g is loxodromic if $\inf_n d_X(x, g^n x)/n > 0$.

3 g is *parabolic* otherwise.

If g is loxodromic then g fixes exactly two points g^+ and g^- of ∂X with north-south dynamics. This means that for any open sets U and V in ∂X containing g^+ and g^- respectively, there exists $N \in \mathbb{N}$ such that for any $n \geq N$, $g^n(\partial X - V) \subset U$. We say that a pair g, h of loxodromic elements is *independent* if $\{g^\pm\} \cap \{h^\pm\} = \varnothing$.

We define the limit set $\Lambda G \subset \partial X$ of the action to be $\overline{Gx} \cap \partial X$. This is independent of the choice of basepoint. We denote by $\mathrm{Fix}_G(\partial X)$ the set of points in ∂X fixed by element of G. With this terminology there is a classification of group actions as follows:

Action	ΛG	Loxodromic elements	Orbits	$\mathrm{Fix}_G\,\partial X$
(1) Elliptic	$\varnothing$	None	Bdd.	Any
(2) Parabolic	1	None	Unbdd.	ΛG
(3) Lineal	2	Yes, and for any loxodromic f and g in G, $\{f^\pm\} = \{g^\pm\}$	Unbdd.	ΛG or $\varnothing$
(4) Quasiparabolic	∞	Yes, and no pair of loxodromic elements is independent	Unbdd.	1
(5) General type	∞	Yes, and there is an independent pair of loxodromic elements	Unbdd.	$\varnothing$

In cases (1)–(3) the action is called *elementary*. For an example of (4), take the action of the Baumslag–Solitar group BS$(1,2)$ on its Bass–Serre tree.

Theorem 3.1.1 *[G] If G acts on X by isometries and X is hyperbolic then exactly one of the possibilities (1)–(5) holds.*

Theorem 3.1.2 *[G] If the action is geometric (i.e. cocompact and properly discontinuous) then exactly one of the possibilities (1), (3) and (5) holds. Moreover, G is finite in case (1) and virtually cyclic in case (3).*

Definition 3.1.3 The action of G on X is *acylindrical* if for any $\epsilon > 0$ there exist R and N such that for any x and y in X with $d_X(x,y) \geq R$, we have

$$|\{g \in G \mid d_X(x, gx) \le \epsilon, d(y, gy) \le \epsilon\}| \le N.$$

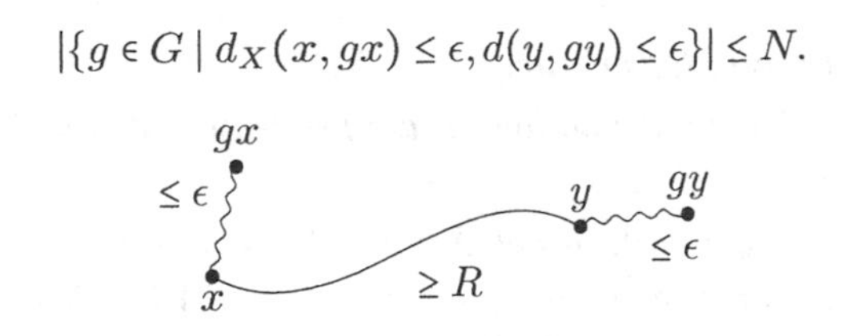

Acylindricity generalises "uniform properness": not all proper actions are acylindrical, but all geometric actions are. The following theorem classifies acylindrical group actions on hyperbolic spaces.

Theorem 3.1.4 *[Os2] If G acts acylindrically on a hyperbolic space X, then exactly one of the following holds.*

1 The action is elliptic.
2 The action is lineal and G is virtually cyclic.
3 The action is of general type.

In particular, an acylindrical action is of general type if and only if it has unbounded orbits and the group is not virtually cyclic.

We recover the following result of Bowditch as a corollary.

Corollary 3.1.5 *[Bo2] If the action of G on X is acylindrical then every element of G is either elliptic or hyperbolic.*

Proof Apply the theorem to the restriction of the action to a cyclic subgroup $\langle g \rangle$. This action cannot be of general type since if a group is of general type then it contains F_2 by a standard ping-pong lemma argument. □

Definition 3.1.6 A group G is called *acylindrically hyperbolic* if G admits a non-elementary acylindrical action on a hyperbolic metric space.

This is usually difficult to check by hand. Fortunately, the notion of a *WPD element* affords us an easier way to verify acylindrical hyperbolicity.

Definition 3.1.7 Let G act by isometries on a space X. An element $g \in G$ satisfies the *weak proper discontinuity* (or WPD) condition if for all x in X, for all $\epsilon > 0$, there exists N such that the set

$$\{h \in G \colon d_X(x, hx) \le \epsilon \text{ and } d_X(g^N x, hg^N x) \le \epsilon\}$$

is finite.

Theorem 3.1.8 *[Os2] For a group G, the following are equivalent:*

1 G is acylindrically hyperbolic.
2 G has a non-elementary action on a hyperbolic space with a loxodromic WPD element.
3 There exists a generating set X for G such that the Cayley graph $\Gamma(G,X)$ is hyperbolic, non-elementary (i.e. $|\partial\Gamma(G,X)| > 2$) and such that the action of G on $\Gamma(G,X)$ is acylindrical.

Finding a WPD element is usually the easiest way to show a group is acylindrically hyperbolic.

We end this section with some examples to illustrate the breadth of the class of acylindrically hyperbolic groups.

Example 3.1.9 Non-elementary hyperbolic and relatively hyperbolic groups are acylindrically hyperbolic.

Example 3.1.10 [Bo2, MM] Let $S_{g,p}$ be the surface of genus g with p punctures. The mapping class group

$$\mathrm{Map}(S_{g,p}) = \mathrm{Homeo}^+(S_{g,p})/\mathrm{Homeo}_0(S_{g,p})$$

of orientation-preserving homeomorphisms of $S_{g,p}$ up to isotopy is acylindrically hyperbolic except when (g,p) is one of $(0,0)$, $(0,1)$, $(0,2)$ and $(0,3)$.

Example 3.1.11 [BF] For a free group F_n with $n \geq 2$ the outer automorphism group $\mathrm{Out}(F_n) = \mathrm{Aut}(F_n)/\mathrm{Inn}(F_n)$ is acylindrically hyperbolic.

Example 3.1.12 [MO, WZ] If G is the fundamental group of a compact irreducible 3-manifold then either G is virtually solvable, the manifold is Seifert fibred, or G is acylindrically hyperbolic.

Example 3.1.13 [Os1] Groups of deficiency at least 2 are acylindrically hyperbolic.

For the next example, we first recall the following definition.

Definition 3.1.14 An isometry g of a CAT(0) space is *rank-one* if it is *axial* (so it acts by translation on a bi-infinite geodesic, called an *axis* of g), and it has an axis that does not bound a flat half-plane. Recall that a *flat half-plane* is a geodesically embedded subspace isometric to a euclidean half-space.

Example 3.1.15 [S] Let G be a group that acts properly on a proper CAT(0) space such that G contains a rank-one isometry and is not virtually cyclic. Then G is acylindrically hyperbolic.

Definition 3.1.16 Let Γ be a finite graph with vertex set $V(\Gamma)$ and edge set $E(\Gamma)$. For an edge e let $o(e)$ be its origin and $t(e)$ be its terminus. The right-angled Artin group $A(\Gamma)$ is defined to be the group with the following presentation:

$$\langle V(\Gamma) \mid [o(e), t(e)] \text{ for each } e \in E(\Gamma)\rangle.$$

Example 3.1.17 [S, CS, Os2] (Sisto, Caprace–Sageev) Any right-angled Artin group that is not either cyclic or a non-trivial direct product is acylindrically hyperbolic.

3.2 Small-cancellation theory

The following "theorem" describes a general small-cancellation philosophy in hyperbolic-like groups; see e.g. [G, H, Ol, Os3]:

"Theorem" 3.2.1 *If G is "hyperbolic-like" and $R_1, \dots, R_k \in G$ are "sufficiently independent" then $G/\langle\langle R_1, \dots, R_k\rangle\rangle$ is "hyperbolic-like", "locally looks like G" and is "as nice as G was".*

To demonstrate this philosophy, we state the following useful theorem.

Theorem 3.2.2 *[H] Let G be an acylindrically hyperbolic group. That is, there exists a generating set X of G such that $\Gamma(G, X)$ is hyperbolic, non-elementary and the action is acylindrical. Let $S \leq G$ be a suitable (defined below) subgroup. Let $T \subset G$ be a finite subset. Then for all $N \in \mathbb{N}$ there exists a homomorphism $\epsilon\colon G \to Q$ such that*

1 Q is acylindrically hyperbolic,
2 $\epsilon|_{\mathrm{Ball_G(1,N)}}$ is injective,
3 $\epsilon(T) \subset \epsilon(S)$ and
4 if G is torsion free then so is Q.

Definition 3.2.3 Let G be an acylindrically hyperbolic group and fix a generating set X for G as in Theorem 3.2.2. Let $S \leq G$. Then S is *suitable* if

1 S is non-elementary (i.e. the restriction of the action of G on $\Gamma(G, X)$ to S is non-elementary) and

2 S does not normalise any finite non-trivial subgroup of G.

For example, if $G = F(a,b,c)$ then $S = \langle a,b \rangle$ is suitable.

Sketch proof of Theorem 3.2.2 If S is suitable then there exist independent loxodromic elements a and b of S. If $T = \{t_1,\dots,t_k\}$ let $R_i = t_i w_i(a,b)$ where $w_i(a,b)$ is a "small-cancellation" word in the alphabet $\{a,b\}$ and let Q be the quotient $G/\langle\langle R_2,\dots,R_k \rangle\rangle$. This definition ensures part (3) of the theorem holds. Part (1) is the "hyperbolic-like" part of the general philosophy. "Locally looks like" gives us (2). (4) is a "nice property" that descends to the quotient. □

We now describe some applications. Our first application is that every acylindrically hyperbolic group is SQ-universal. Let us first recall the definition of this property.

Definition 3.2.4 A group G is *SQ-universal* if every countable group is a subgroup of a quotient of G.

Remark 3.2.5 If G is SQ-universal, then $F_2 \le G$. If in addition G is countable then it has $2^{\aleph_0}$ normal subgroups since there are $2^{\aleph_0}$ distinct finitely generated groups.

It is sufficient to replace "countable" by "finitely generated" in the definition of SQ-universality:

Theorem 3.2.6 *[GNN] Every countable group can be embedded in a 2-generator group, so F_2 is SQ-universal.*

SQ-universality is known for some groups of geometric origin, such as hyperbolic groups. The following theorem generalises such results, and provides new information about some other classical groups, such as mapping class groups.

Theorem 3.2.7 *[DGO] Any acylindrically hyperbolic group A is SQ-universal.*

Proof Let a given finitely generated group H have generators $t_1,\dots,t_k$. Then $G = A * H$ is acylindrically hyperbolic and A is suitable in G. Apply Theorem 3.2.2 in the case $N = 1$ to G with $S = A$ and $T = \{t_1,\dots,t_k\}$, assuming without loss of generality that the generating set X contains H. Then $Q = \langle \epsilon(T), \epsilon(A) \rangle = \langle \epsilon(A) \rangle$ is a quotient of A. The subgroup H is contained in the ball $\mathrm{Ball}_G(1,1)$, so $\epsilon|_H$ is injective, i.e. $H \hookrightarrow Q$. □

Using the following theorem we deduce a corollary about mapping class groups.

Theorem 3.2.8 *[DGO] If $G \leq \mathrm{Map}(\mathrm{S_{g,p}})$ then either G is virtually abelian or G surjects onto an acylindrically hyperbolic group.*

Corollary 3.2.9 *If $G \leq \mathrm{Map}(\mathrm{S_{g,p}})$ then G is either virtually abelian or SQ-universal.*

This implies:

1 *The Tits alternative for subgroups of* $\mathrm{Map}(\mathrm{S_{g,p}})$ *[M, I].*
2 *Any homomorphism $SL_3\mathbb{Z} \to \mathrm{Map}(\mathrm{S_{g,p}})$ has finite image [FM].*

We can also use this small-cancellation theory for acylindrically hyperbolic groups to answer the following natural question:

Question 3.2.10 Does there exist a finitely generated group, other than $\{1\}$ and $\mathbb{Z}/2\mathbb{Z}$, in which all non-trivial elements are conjugate?

Remark 3.2.11 Such a group must be torsion-free. Therefore G cannot be a limit of hyperbolic groups:

$$G_0 \twoheadrightarrow G_1 \twoheadrightarrow G_2 \twoheadrightarrow \dots \twoheadrightarrow G_\infty = G \ni a, t \text{ such that } t^{-1}at = a^2$$

The relation must hold in some G_i, which is impossible, since a has infinite order.

Therefore, to find a positive answer to Question 3.2.10, we must go *beyond hyperbolicity.*

Theorem 3.2.12 *The answer to the question is yes!*

This result was originally proved in [Os3]. Here we sketch the later proof given in [H], in which the result is presented as an application of small cancellation theory in acylindrically hyperbolic groups.

Proof $G_0 = F(a,b)$ is acylindrically hyperbolic. Let $G_0 = \{1, g_0, g_1, \dots\}$. Assume that G_n is constructed and satisfies the following inductive hypotheses:

1 G_n is acylindrically hyperbolic.
2 Non-trivial elements from $\{g_0, \dots, g_n\}$ are conjugate in G_n.
3 G_n is torsion free.

Then let $G_{n+1/2} = \langle G_n, t \mid t^{-1} g_{n+1} t = g_0 \rangle$ if g_{n+1} is non-trivial in G_n and $G_{n+1/2} = G_n$ otherwise. By [H, Prop 6.2] this group is acylindrically hyperbolic and G_n is suitable in $G_{n+1/2}$. Let G_{n+1} be the group Q given by Theorem 3.2.2 applied with $G = G_{n+1/2}$, $S = G_n$ and $T = \{t\}$. Let

$G = G_\infty$; this has exactly two conjugacy classes. If it were finite it would be finitely presented, and therefore it would have to be G_n for some n. □

3.3 Dehn surgery

We first recall the classical definition of Dehn surgery on 3-manifolds. Let M be a closed 3-manifold and let K be an embedded knot in M. Let $N = M - N(K)$ where $N(K)$ is an open regular neighbourhood of K. Now choose a homeomorphism $\phi: \partial N \to \partial T^2$, where $T^2 = S^1 \times D^2$. Then the space $N \cup_\phi T^2$ is determined up to homeomorphism by the element $\sigma = [\phi(\partial D^2)] \in \pi_1 \partial N$; this element is called the *slope*. Let $N(\sigma) = N \cup_\phi T^2$. Then $\pi_1(N(\sigma)) = \pi_1 N / \langle\langle \sigma \rangle\rangle^{\pi_1 N}$.

Theorem 3.3.1 *[T] Suppose that N has a complete finite volume hyperbolic metric. Then $N(\sigma)$ is hyperbolic for all but finitely many $\sigma \in \pi_1 \partial N$.*

Corollary 3.3.2 *$\pi_1 N / \langle\langle \sigma \rangle\rangle^{\pi_1 N}$ is hyperbolic for all but finitely many $\sigma \in \pi_1 \partial N$.*

We now aim to describe a similar sort of surgery for acylindrically hyperbolic groups. This theory is related to classical Dehn surgery by the following correspondence.

Complete finite volume hyperbolic manifold N	$\longleftrightarrow$	Acylindrically hyperbolic group
∂N	$\longleftrightarrow$	Hyperbolically embedded subgroup H
Slope $\sigma \in \pi_1 \partial N$	$\longleftrightarrow$	$N \trianglelefteq H$

In order to describe surgery for acylindrically hyperbolic groups we must first define hyperbolically embedded subgroups.

Definition 3.3.3 Let G be a group and let $X \subset G$. Then we define $\Gamma(G, X)$ to be the graph with vertex set G and an edge joining g to gx for each $g \in G$ and $x \in X$. Note that when X is a generating set for G, this is the Cayley graph of G with respect to X.

Definition 3.3.4 [DGO] Let G be a group with $X \subset G$ and $H \leq G$ such that $G = \langle X \cup H \rangle$. Let $\Gamma = \Gamma(G, X \cup H)$; this space contains $\Gamma(H, H)$, which we shall denote Γ_H. Let $\Delta = \Gamma - E(\Gamma_H)$. For $g, h \in H$, let $d_H(g, h)$ be the minimal length of a path from g to h in Δ, or ∞ if no such path exists. We say that H is *hyperbolically embedded* in G with respect to X (and write $H \hookrightarrow_h (G, X)$) if:

1 Γ is hyperbolic, and
2 For any $n \in \mathbb{N}$ there are only finitely many elements $h \in H$ such that $d_H(1, h) \leq n$.

We write $H \hookrightarrow_h G$ if there exists X such that $H \hookrightarrow_h (G, X)$.

Example 3.3.5 $G \hookrightarrow_h G$ by taking $X = \varnothing$. If $|H| < \infty$ then $H \hookrightarrow_h G$ by taking $X = G$. These examples are called *degenerate* and all other examples are *non-degenerate.*

Example 3.3.6 Let $G = \mathbb{Z} \times H$ with the $\mathbb{Z}$ factor generated by an element x. Let $X = \{x\}$. Then Γ is:

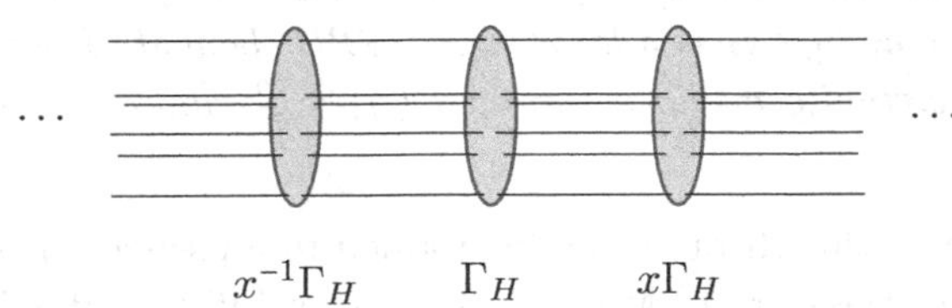

This is quasi-isometric to a line, so is hyperbolic. However, for any $g \neq h \in H$, $d_H(g, h) = 3$, so condition (2) fails as long as H is infinite.

By a similar argument, one can prove the following theorem, which provides an obstruction to a subgroup being hyperbolically embedded.

Theorem 3.3.7 *[DGO] If $H \hookrightarrow_h G$ then H is almost-malnormal in G. That is, $|gHg^{-1} \cap H| < \infty$ for any $g \in G - H$.*

Corollary 3.3.8 *If $H \hookrightarrow_h G$ and $|G \colon H| < \infty$ then either G is finite or $H = G$.*

Example 3.3.9 $G = \mathbb{Z} * H$, $X = \{x\}$. Then Γ is a "tree" in which vertices are copies of Γ_H.

Then $d_H(h_1, h_2) = \infty$ as for any $h_1 \neq h_2$ in H, so $H \hookrightarrow_h G$.

The definition of a hyperbolically embedded subgroup is intended to generalise the notion of relative hyperbolicity:

Theorem 3.3.10 *[DGO] A group G is hyperbolic relative to $H \leq G$ if and only if $H \hookrightarrow_h (G, X)$ for some finite subset $X \subset G$.*

This can be generalised to the case of a collection of subgroups, but in this course we work with only one subgroup at a time to simplify notation.

It follows from the theorem that if we restrict to the case in which the subset X is finite, we get nothing new beyond the theory of relatively hyperbolic groups: for interesting new examples we must allow X to be infinite. Before giving new examples of the property we note the following theorem relating this definition to acylindrical hyperbolicity.

Theorem 3.3.11 *[DGO, Os2] A group G is acylindrically hyperbolic if and only if G has a non-degenerate hyperbolically embedded subgroup.*

We might now ask what sort of hyperbolically embedded subgroups an acylindrically hyperbolic group can have.

Theorem 3.3.12 *[DGO] Suppose that G acts acylindrically on a hyperbolic space and $g \in G$ is a loxodromic WPD element. Then there exists a maximal virtually cyclic subgroup $E(g) \leq G$ containing g such that $E(g) \hookrightarrow_h G$.*

For example, this theorem can be applied to a pseudo-Anosov element of a mapping class group $\mathrm{Map}(\mathrm{S}_{\mathrm{g,p}})$, or to a fully irreducible element of $\mathrm{Out}(F_n)$. Note that these groups are not hyperbolic relative to any proper subgroup.

Theorem 3.3.12 shows that acylindrically hyperbolic groups contain many virtually cyclic hyperbolically embedded subgroups. The following theorem shows that they also contain free hyperbolically embedded subgroups.

Theorem 3.3.13 *[DGO] If G is an acylindrically hyperbolic group, then for every $n \in \mathbb{N}$ there is a finite group K such that $F_n \times K \hookrightarrow_h G$.*

We can now state the main theorem about surgery on acylindrically hyperbolic groups.

Theorem 3.3.14 *[DGO] Suppose that G is an acylindrically hyperbolic group and $H \hookrightarrow_h (G, X)$. Then there exists a finite subset $F \subset H - \{1\}$ such that for every normal subgroup $N \trianglelefteq H$ satisfying $N \cap F = \emptyset$ we have:*

1 $\langle\langle N \rangle\rangle^G \cap H = N$. Equivalently $H/N \to \bar{G}$, where $\bar{G} = G/\langle\langle N \rangle\rangle^G$, is injective, so we may consider H/N as a subgroup of $\bar{G}$.
2 $H/N \hookrightarrow_h (\bar{G}, \bar{X})$, where $\bar{X} = X\langle\langle N \rangle\rangle^G / \langle\langle N \rangle\rangle^G$.
3 $\langle\langle N \rangle\rangle^G = \ast_{t \in T}\, tNt^{-1}$ for some $T \subset G$.
4 If $N_1 \trianglerighteq N_2 \trianglerighteq N_3 \trianglerighteq \cdots$ and $\bigcap N_i = 1$ then $\bigcap \langle\langle N_i \rangle\rangle^G = 1$.

A general philosophy for hyperbolically embedded subgroups is that they behave like free factors; note that each of the parts of Theorem 3.3.14 is easy to prove in this case.

To justify the description of Theorem 3.3.14 as "group-theoretic Dehn surgery", let us consider the following special case. If G is hyperbolic relative to H, then the theorem tells us that $G/\langle\langle N\rangle\rangle^G$ is hyperbolic relative to H/N. In the context of classical Dehn surgery, if $G = \pi_1 N$ and $H = \pi_1(\partial N)$, then the assumption that N admits a finite volume complete hyperbolic metric ensures that G is hyperbolic relative to H. Then the theorem says that for all but finitely many $\sigma \in \pi_1 \partial N$ the group $\pi_1(N(\sigma))$ is hyperbolic relative to $H/\langle\sigma\rangle$. But $H/\langle\sigma\rangle$ is virtually cyclic, so this means that $\pi_1(N(\sigma))$ is hyperbolic. Thus the group-theoretic version of Thurston's theorem is recovered.

We also deduce the following corollary.

Corollary 3.3.15 *Any mapping class group* $\mathrm{Map}(\mathrm{S_{g,p}})$ *with* $3g + p > 4$ *contains a normal non-trivial purely pseudo-Anosov subgroup.*

Proof Let g be pseudo-Anosov. Then $E(g) \hookrightarrow_h G$. Without loss of generality we may assume that $\langle g\rangle \trianglelefteq E(g)$. Then applying the theorem we obtain $N = \langle g^m\rangle \trianglelefteq E(g)$, and $\langle\langle g^m\rangle\rangle^G$ is free. This subgroup can be shown to be purely pseudo-Anosov: in general, in the setting of Theorem 3.3.14, for any $g \in \langle\langle N\rangle\rangle^G$, either g acts loxodromically on $\Gamma(G, X \cup H)$ or g is conjugate to an element of N. □

3.4 The extension problem

Let $H \leq G$ and suppose that H acts on a metric space S. It is then natural to ask the following question.

Question 3.4.1 When can you extend this action to G?

Definition 3.4.2 An action of a group G on a metric space R is an *extension* of the action of H on S if there exists a coarsely H-equivariant quasi-isometric embedding $f\colon S \to R$. More precisely:

1 For every $s \in S$, $\sup_{h\in H} d_R(f(hs), hf(s)) < \infty$.
2 There exists $C > 0$ such that for all x and y in S,

$$\frac{1}{C} d_S(x,y) - C \leq d_R(f(x), f(y)) \leq C d_S(x,y) + C.$$

Example 3.4.3 If H acts on S with bounded orbits, then the trivial action of G on S is an extension.

The *extension problem* for the action of $H \leq G$ on S is the question of the existence of an action of G on a metric space R that is an extension of the action of H on S. The extension problem is *solvable* for $H \leq G$ if the answer to this question is "yes" for all actions of H on metric spaces.

Example 3.4.4 If $G = \mathrm{Sym}(\mathbb{N})$ then every countable group embeds into G, but every action of G on a metric space has bounded orbits [Be, dC]. Thus if H is any group that admits an action on a metric space with unbounded orbits, then the extension problem is not solvable for $H \leq G$.

One might also want a countable non-example. First recall the following easy lemma, which is one of the steps in the proof of the Švarc–Milnor lemma.

Lemma 3.4.5 *Let G be a finitely generated group. Then for any action of G on a metric space R, the orbit map $g \mapsto gr$, for any $r \in R$, is a Lipschitz map $G \to R$.*

Example 3.4.6 Let G be the Baumslag–Solitar group

$$\mathrm{BS}(1,2) = \langle a, b \mid b^{-1}ab = a^2 \rangle$$

and let $H = \langle a \rangle$. Then H acts on $\mathbb{R}$ by translation. We claim that this action does not extend to an action of G. Suppose it does and $f : \mathbb{R} \to R$ is as in the definition of an extension. Then we have the following inequality:

$$\begin{aligned} d_R\left(f(r), f(a^{2^m} r)\right) &\geq \frac{1}{C} d_{\mathbb{R}}\left(r, a^{2^m} r\right) - C \\ &= 2^m/C - C. \end{aligned}$$

But for some C',

$$\begin{aligned} d_R\left(f(r), f(a^{2^m} r)\right) &\leq d_R\left(f(r), a^{2^m} f(r)\right) + C' \\ &\leq C'' |a^{2^m}|_{\{a,b\}} + C' \\ &\leq C''(2m+1) + C'. \end{aligned}$$

This is a contradiction. The problem here was that $\langle a \rangle$ is distorted in $\mathrm{BS}(1,2)$.

This example is one incarnation of a general obstruction to the solvability of the extension problem, as the following proposition shows. We leave the proof as an exercise.

Proposition 3.4.7 *If G is finitely generated and the extension problem is solvable for $H \leq G$ then H is finitely generated and undistorted in G.*

This proposition does not give a sufficient condition for the solvability of the extension problem:

Example 3.4.8 Let $G = \langle a, b, t \mid t^2 = 1, t^{-1}at = b \rangle$ and let $H = \langle a, b \rangle$. Let H act on $\mathbb{R}$ by

$$a\colon x \to x + 1$$

$$b\colon x \to x.$$

This action does not extend. To see this, note that in the action of H we have that a has unbounded orbits, while b does not. However, in G the elements a and b are conjugate, so one must have unbounded orbits if and only if the other does.

However, H has finite index in G, so is certainly finitely generated and undistorted.

It is difficult to see how to formalise the problem with this example. One might suggest that H must be almost-malnormal in G for the extension problem to be solvable, but this is not the case, as the following example shows.

Example 3.4.9 If H is a retract of G, then the extension problem for $H \leq G$ is solvable. To see this, let $\rho\colon G \to H$ be the retraction and let S be a space on which H acts. We define $gs = \rho(g)s$ for each g in G and s in S.

The following theorem gives a sufficient condition.

Theorem 3.4.10 *[AHO] If $H \hookrightarrow_h G$ then the extension problem for $H \leq G$ is solvable.*

We sketch the proof.

Sketch proof Let Γ and Γ_H be as in the definition of a hyperbolically embedded subgroup. Then glue a copy of the space S on which H acts to each translate $g\Gamma_H$ of Γ_H in Γ. That H is hyperbolically embedded ensures that the inclusion of S into this space is a quasi-isometric embedding. □

The class of hyperbolically embedded subgroups is quite general among the class of all subgroups with solvable extension problem. For example, we can prove the following corollary for hyperbolic groups.

Corollary 3.4.11 *If G is hyperbolic and $H \leq G$ has solvable extension problem, then H is quasi-convex. Conversely, if H is quasi-convex and almost-malnormal, then the extension problem for $H \leq G$ is solvable.*

Proof If $H \leq G$ has solvable extension problem, then it is finitely generated and undistorted, so it is quasi-convex since G is hyperbolic.

Conversely, if H is quasi-convex and almost-malnormal, then G is hyperbolic relative to H [Bo1], so $H \hookrightarrow_h G$, which implies that the extension problem is solvable by Theorem 3.4.10 □

Finally, we note that almost-malnormality is not necessary:

Example 3.4.12 Let $G = F(a,b)$ and let $H = \langle a^2 \rangle$. Then the extension problem for $H \leq G$ is solvable, but H is not almost-malnormal in G.

3.5 Acylindrically hyperbolic structures

The study of the extension problem has an application to the study of acylindrically hyperbolic groups. In this section we explore this further.

Take cobounded actions of G on spaces S and R; denote these actions by A and B. We write $A \leq B$ if there exists a coarsely G-equivariant Lipschitz map $f\colon R \to S$. We say that $A \sim B$ if $A \leq B$ and $B \leq A$. We denote by $[A]$ the set of B such that $A \sim B$.

Definition 3.5.1 An *acylindrically hyperbolic structure* on a group G is an equivalence class of cobounded acylindrical actions of G on hyperbolic spaces. Let $\mathcal{AH}(G)$ be the set of all acylindrically hyperbolic structures on G. (If *all* metric spaces are allowed, then $\mathcal{AH}(G)$ might not be a set; this problem is fixed by restricting to metric spaces of cardinality at most that of the continuum.) This set is partially ordered by $\leq$.

Note that the trivial action of G on a point is always in $\mathcal{AH}$. In general we have the following trichotomy.

Theorem 3.5.2 *[ABO] For every group G, exactly one of the following holds:*

1 $|\mathcal{AH}(G)| = 1$.
2 $|\mathcal{AH}(G)| = 2$ *and G is virtually cyclic.*
3 $|\mathcal{AH}(G)| = \infty$ *and G is acylindrically hyperbolic.*

It is now natural to try to understand the order structure on $\mathcal{AH}(G)$ in the case when G is acylindrically hyperbolic. For example, one might ask whether or not it has a maximal element. (It certainly has a minimal element: the trivial action on a point.) In fact, some acylindrically hyperbolic groups do have maximal acylindrically hyperbolic structure, such as hyperbolic groups and mapping class groups, and some do not. It is not known whether or not an acylindrically hyperbolic group without a maximal acylindrically hyperbolic structure can be finitely presented.

After considering the question of the existence of maximal and minimal elements of $\mathcal{AH}(G)$, one might wish to know the possible lengths of chains and antichains in the set. (Recall that a *chain* is a totally ordered subset and an *antichain* is a subset in which any two distinct elements are incomparable.)

Theorem 3.5.3 *[ABO] If G is acylindrically hyperbolic, then $\mathcal{AH}(G)$ has chains and antichains of cardinality $2^{\aleph_0}$.*

These chains and antichains are of maximal possible cardinality: if G is countable then one can show that $\mathcal{AH}(G)$ has size at most $2^{\aleph_0}$.

The proof of this theorem uses Theorem 3.4.10.

Sketch proof First one shows that the theorem holds if $G \cong F_2$.

There exists a subgroup H isomorphic to the direct product of F_2 by a finite group such that $H \hookrightarrow_h G$. Then the extension problem for $F_2 \leq H$ is solvable since it is a retract, and the extension problem for $H \leq G$ is solvable by Theorem 3.4.10. Moreover, one can add to the statement of Theorem 3.4.10 that, in the notation of the theorem, if the action of H on S is acylindrical and S is hyperbolic, then the same is true of the action of G.

It follows that any acylindrically hyperbolic structure on F_2 can be lifted to an acylindrically hyperbolic structure on G; in other words, we obtain $\mathcal{AH}(F_2)$ is a retract of $\mathcal{AH}(G)$. The theorem follows. □

References

[ABO] C. Abbott, S. Balasubramanya, and D. Osin. Hyperbolic structures on groups. In preparation.

[AHO] C. Abbott, D. Hume, and D. Osin. Extending group actions on metric spaces. *ArXiv e-prints*, March 2017.

[Be] George M. Bergman. Generating infinite symmetric groups. *Bull. London Math. Soc.*, 38(3):429–440, 2006.

[BF] Mladen Bestvina and Mark Feighn. A hyperbolic $\mathrm{Out}(F_n)$-complex. *Groups Geom. Dyn.*, 4(1):31–58, 2010.

[Bo1] B.H. Bowditch. Relatively hyperbolic groups. *Internat. J. Algebra Comput.*, 22(3):1250016, 66, 2012.

[Bo2] Brian H. Bowditch. Tight geodesics in the curve complex. *Invent. Math.*, 171(2):281–300, 2008.

[CS] Pierre-Emmanuel Caprace and Michah Sageev. Rank rigidity for CAT(0) cube complexes. *Geom. Funct. Anal.*, 21(4):851–891, 2011.

[DGO] F. Dahmani, V. Guirardel, and D. Osin. Hyperbolically embedded subgroups and rotating families in groups acting on hyperbolic spaces. *Mem. Amer. Math. Soc.*, 245(1156):v+152, 2017.

[dC] Yves de Cornulier. Strongly bounded groups and infinite powers of finite groups. *Comm. Algebra*, 34(7):2337–2345, 2006.

[FM] Benson Farb and Howard Masur. Superrigidity and mapping class groups. *Topology*, 37(6):1169–1176, 1998.

[G] M. Gromov. Hyperbolic groups. In *Essays in group theory*, volume 8 of *Math. Sci. Res. Inst. Publ.*, pages 75–263. Springer, New York, 1987.

[GNN] Graham Higman, B.H. Neumann, and Hanna Neumann. Embedding theorems for groups. *J. London Math. Soc.*, 24:247–254, 1949.

[H] Michael Hull. Small cancellation in acylindrically hyperbolic groups. *Groups Geom. Dyn.*, 10(4):1077–1119, 2016.

[I] N.V. Ivanov. Algebraic properties of the Teichmüller modular group. *Dokl. Akad. Nauk SSSR*, 275(4):786–789, 1984.

[MM] Howard A. Masur and Yair N. Minsky. Geometry of the complex of curves. I. Hyperbolicity. *Invent. Math.*, 138(1):103–149, 1999.

[M] John McCarthy. A "Tits-alternative" for subgroups of surface mapping class groups. *Trans. Amer. Math. Soc.*, 291(2):583–612, 1985.

[MO] Ashot Minasyan and Denis Osin. Acylindrical hyperbolicity of groups acting on trees. *Math. Ann.*, 362(3-4):1055–1105, 2015.

[Ol] A. Yu. Ol'shanskiĭ. On residualing homomorphisms and G-subgroups of hyperbolic groups. *Internat. J. Algebra Comput.*, 3(4):365–409, 1993.

[Os1] D. Osin. On acylindrical hyperbolicity of groups with positive first ℓ^2-Betti number. *Bull. Lond. Math. Soc.*, 47(5):725–730, 2015.

[Os2] D. Osin. Acylindrically hyperbolic groups. *Trans. Amer. Math. Soc.*, 368(2):851–888, 2016.

[Os3] Denis Osin. Small cancellations over relatively hyperbolic groups and embedding theorems. *Ann. of Math. (2)*, 172(1):1–39, 2010.

[S] A. Sisto. Contracting elements and random walks. *Journal für die reine und angewandte Mathematik*, to appear.

[T] W. P. Thurston. *The geometry and topology of three-manifolds.* Princeton lecture notes, 1979.

[WZ] Henry Wilton and Pavel Zalesskii. Profinite properties of graph manifolds. *Geom. Dedicata*, 147:29–45, 2010.

PART TWO

EXPOSITORY ARTICLES

4

A survey on Morse boundaries and stability

Matthew Cordes

Department of Mathematics
Technion - IIT
Haifa 32000
Israel

4.1 Generalizing hyperbolicity

Recently there have been efforts to generalize tools used in the setting of Gromov hyperbolic spaces to larger classes of spaces. We survey here two aspects of these efforts: Morse boundaries and stable subspaces. Both the Morse boundary and stable subspaces are systematic approaches to collect and study the hyperbolic aspects of finitely generated groups. This survey will explain a nice relationship between the two approaches.

Throughout the survey we will expect the reader to be familiar with the basics of geometric group theory. For details on this material see [BH] and for additional information especially on boundaries of not-necessarily-geodesic hyperbolic spaces see [BS].

We begin with an important definition that has its roots in a classical paper of Morse [Mo]:

Definition 4.1.1 A geodesic γ in a metric space is called *N-Morse* if there exists a function $N = N(\lambda, \epsilon)$ such that for any (λ, ϵ)-quasi-geodesic σ with endpoints on γ, we have $\sigma \subset \mathcal{N}_N(\gamma)$, the N-neighbourhood of γ. We call the function $N\colon \mathbb{R}_{\geq 1} \times \mathbb{R}_{\geq 0} \to \mathbb{R}_{\geq 0}$ a *Morse gauge.* A geodesic is *Morse* if it is N-Morse for some N.

In a geodesic δ-hyperbolic space, the Morse lemma says there is a Morse gauge N which depends only on δ such that every ray is N-Morse, i.e., the behaviour of quasi-geodesics (on a large scale) is similar to that of geodesics. This property fails in a space like $\mathbb{R}^2$. On the other hand, if every geodesic in some geodesic space is N-Morse, then the space is δ-hyperbolic, where δ depends on N.

There are a few other competing definitions of geodesics that admit "hyperbolic like" properties: geodesics that satisfy a contracting property, geodesics with superlinear divergence, and having cut points in the asymptotic cone. (See Section 4.2.1 for the definition of the contracting property and [CS2, ACGH1, DS] for the others.) All of these notions have been used extensively to analyse many groups and spaces: right-angled Artin groups [BC, KMT, CH2], Teichmüller space [Mins, Be, Su, BrF, BMM], the mapping class group and curve complex [MM1, MM2, DT, BBF], CAT(0) spaces [BD, Su], Out(F_n) [A-K], relatively hyperbolic groups and spaces [DS, Osi2], acylindrically hyperbolic groups [DGO, Si, BBF], and small cancellation groups [ACGH2], among others. Furthermore these properties find applications in rigidity theorems such as Mostow Rigidity in rank-1 [Pau] and the Rank Rigidity Conjecture for CAT(0) spaces [BaB, BeF, CF].

Geodesics of these types all have a relationship to the Morse property. In [CS2], Charney and Sultan show that Morse, superlinear 'lower divergence', and 'strongly' contracting are equivalent notions in CAT(0) spaces. In [ACGH1] the authors characterize Morse quasi-geodesics in arbitrary geodesic metric spaces and show they are 'sublinearly' contracting and have 'completely superlinear' divergence. In addition to several other characterizations of Morse geodesics, the authors in [DMS] show that a quasi-geodesic γ is Morse if and only if for every asymptotic cone $\mathcal{C}$ and every point x on the limit $\bar{\gamma}$ in $\mathcal{C}$, the two connected parts of $\bar{\gamma} \smallsetminus \{x\}$ are in two different connected components of $\mathcal{C} \smallsetminus \{x\}$.

While groups may have no infinite-length Morse geodesics, Sisto in [Si] shows that every acylindrically hyperbolic group has a bi-infinite Morse geodesic. This class of groups, recently unified by Osin in [Osi3], encompasses many groups of significant interest in geometric group theory: hyperbolic and relatively hyperbolic groups, non-directly decomposable right-angled Artin groups, mapping class groups, and Out(F_n). There are groups outside this class that contain Morse geodesics. Examples of these groups appear in [OOS]; some of these groups have been shown to have infinitely many Morse geodesics but are not acylindrically hyperbolic. At another extreme, any groups that admit a law do not contain any Morse geodesics [DS]. It is easy to see that $\mathbb{R}^n$ for $n \geq 2$ has no Morse geodesics: take any geodesic ray γ and consider a right-angled isosceles triangle whose hypotenuse is on γ. It is easy to check that the other two edges form a $(\sqrt{2}, 0)$-quasi-geodesic with endpoints on γ. This is true no matter how large the triangle and thus these uniform quasi-geodesics can be arbitrarily far from γ and therefore violate the Morse condition.

In this article we will survey notions of Morse boundaries of increasing generality in Sections 4.2 and 4.3. In Section 4.3 we will also introduce stability, the notion of stable equivalence, and two 'stable equivalence' invariants. We end with some calculations of these invariants and some discussion about distinguishing relatively hyperbolic groups and the Bowditch boundary. In Section 4.4 we will spend some time discussing stable subgroups and introduce a boundary characterization of convex cocompactness that is equivalent to stability. Finally, in Section 4.5 we will discuss a topology on the contracting boundary that is second countable and thus metrisable.

4.2 Contracting and Morse boundaries

Boundaries have been an extremely fruitful tool in the study of hyperbolic groups. The visual boundary, as a set, is made up of equivalence classes of geodesic rays, where one ray is equivalent to the other if they fellow travel. Roughly, one topologises the boundary by declaring open neighbourhoods of a ray γ to be the rays that stay close to γ for a long time. Gromov showed that a quasi-isometry between hyperbolic metric spaces X and Y induces a homeomorphism on the visual boundaries. In the setting of a finitely generated group G acting geometrically on X and Y, the quasi-isometry from X to Y induced by these actions extends G-equivariantly to a homeomorphism of their boundaries. This means, in particular, that the boundary of a hyperbolic group (as a topological space) is independent of the choice of (finite) generating set.

The boundary of a hyperbolic group is a powerful tool to study the structure of the group. For instance, Bowditch and Swarup relate topological properties of the boundary to the JSJ decomposition of a hyperbolic group [Bo1, Sw]. Bestvina and Mess in [BM] relate the virtual cohomological dimension of a hyperbolic group G to the dimension of the boundary of G. Boundaries have also been instrumental in the proofs of quasi-isometric rigidity theorems, particularly Mostow Rigidity in rank 1 [Pau].

There is also a robust boundary theory of relatively hyperbolic groups using the Bowditch boundary introduced in a 1998 preprint of Bowditch that was recently published [Bo3]. Bowditch used this boundary to analyse JSJ decompositions [Bo1, Bo2]. Groff showed that if a group G is hyperbolic relative to a collection $\mathcal{A}$ of subgroups that are not properly relatively hyperbolic, then the Bowditch boundary is a quasi-isometry invariant [Grof].

The topology on the visual boundary for a CAT(0) space can be defined in a manner similar to the visual boundary of a hyperbolic space. Hruska and Kleiner show that in the case of a CAT(0) space with isolated flats, this boundary is a quasi-isometry invariant [HK]. Unfortunately, Croke and Kleiner produced an example of a right-angled Artin group which acts geometrically on two quasi-isometric CAT(0) spaces which have non-homeomorphic visual boundaries [CK], hence the boundary of a CAT(0) group is not well-defined. More surprising is that Wilson shows that this group admits an uncountable collection of distinct boundaries [Wi]. Charney and Sultan in [CS1] showed that if one restricts attention to rays with hyperbolic-like behaviour, *contracting rays*, then one can construct a quasi-isometry invariant boundary for any complete CAT(0) space. They call this boundary the *contracting boundary.*

4.2.1 Contracting boundaries

We begin with a description of the contracting boundary of a CAT(0) space introduced in [CS2]. We will start with some explication on contracting geodesics.

Definition 4.2.1 (contracting geodesics) Given a fixed constant D, a geodesic γ is said to be *D-contracting* if for all $x, y \in X$,

$$d_X(x,y) < d_X(x,\pi_\gamma(x)) \implies \operatorname{diam}(\pi_\gamma(x),\pi_\gamma(y)) < D$$

where π_γ is the closest point projection to γ. We say that γ is *contracting* if it is D-contracting for some D. An equivalent definition is that any metric ball B not intersecting γ projects to a segment of length $< 2D$ on γ.

Remark 4.2.2 This definition is sometimes referred to as strongly contracting. To see a more general definition of contracting and its relationship with the Morse property and divergence see Section 4.5 and for more details [ACGH1]. This more general definition is used by Cashen to answer a question of Charney–Sultan (see Theorem 4.2.11) and by Cashen–Mackay to construct a topology on the contracting boundary that is metrisable (see Section 4.5).

It is easy to see that any geodesic in a flat will not be contracting because the projection of any ball onto a geodesic will just be the diameter of the ball. In a hyperbolic CAT(0) space all geodesics are uniformly contracting. Contracting geodesics appear in many groups and spaces,

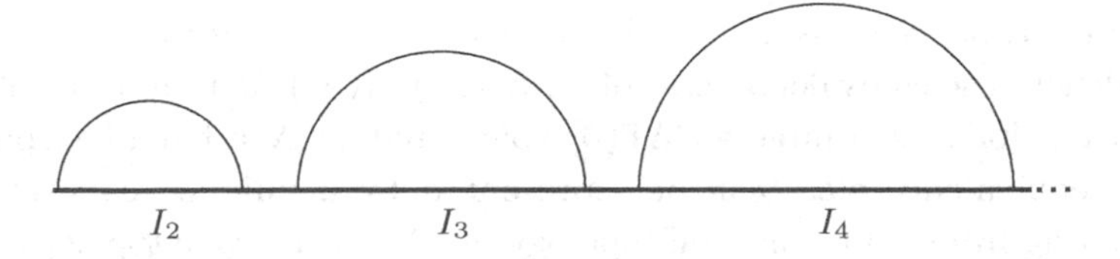

Figure 4.1 Morse and strongly contracting are not equivalent. See Remark 4.2.3.

for instance: pseudo-Anosov axes in Teichmüller space [Mins], iwip axes in the Outer Space of outer automorphisms of a free group [A-K], and axes of rank 1 isometries of CAT(0) spaces [BaB, BeF].

Remark 4.2.3 (Morse and (strongly) contracting are not equivalent) If X is a proper geodesic space, then contracting implies Morse [A-K]. In this generality the converse is not true. Consider a ray Y and a set of disjoint intervals $\{I_i\}_{i\in\mathbb{N}}$ of length i on Y. To the endpoints of these intervals attach an edge of length i^2. Let X be the resulting space. See Figure 4.1. The ray $Y \subset X$ is not contracting because balls can have unbounded projection onto Y. To see that Y is Morse, we need to check that there exits a constant $N = N(\lambda, \epsilon)$ so that for any (λ, ϵ)-quasi-geodesic σ with endpoints on Y, $\sigma \subset \mathcal{N}_N(Y)$. By [BH, III.H Lemma 1.11] we can replace σ with a 'tame' quasi-geodesic σ' with three nice properties: σ and σ' have the same endpoints, σ and σ' have Hausdorff distance less than $(\lambda + \epsilon)$, and the length between any two points x, y on σ' is bounded by a linear function of $d(x, y)$ (with constants depending only on λ, ϵ). Since the length of the attached intervals is i^2 while the distance between their endpoints is i, in order for σ' to cross such a segment, we must have i^2 less than a linear function of i. Thus σ' must only cross finitely many of the attached segments. Thus for any (λ, ϵ)-quasi-geodesic with endpoints on Y we can choose $N(\lambda, \epsilon) = j^2 + \lambda + \epsilon$ where j^2 is the length of the longest attached interval σ' travels over.

In the setting of CAT(0) spaces, Sultan shows that Morse and contracting are equivalent notions [Su] and in [CS2] Charney and Sultan reprove this fact with explicit control on the constants. Charney and Sultan also characterize contracting geodesics in CAT(0) cube complexes using a combinatorial criterion that gives an effective tool for analysing the boundary. We will use this characterization to help us compute Examples 4.2.6 and 4.2.8.

Let X be a CAT(0) space. We define the visual boundary ∂X to be the set of equivalence classes of geodesic rays up to asymptotic equivalence and denote the equivalence class of a ray α by $\alpha(\infty)$. It is an elementary fact that, for X a complete CAT(0) space and $e \in X$ a fixed basepoint, every equivalence class can be represented by a unique geodesic ray emanating from e. One natural topology on ∂X is the cone topology. We define the topology of the boundary with a system of neighbourhood bases. A neighbourhood basis for $\alpha(\infty)$ is given by open sets of the form:

$$\begin{aligned} &U(\alpha, r, \epsilon) \\ &\quad = \{\beta(\infty) \in \partial X \mid \beta \text{ a ray based at } e \text{ and } \forall t < r, d(\beta(t), \alpha(t)) < \epsilon\}. \end{aligned}$$

That is, two geodesic rays in the cone topology are close if they fellow travel (at distance less than ϵ) for a long time (at least time r). This topology is independent of choice of basepoint. When we refer to the visual boundary we will always mean with the cone topology, unless otherwise stated.

Let X be a complete CAT(0) space with basepoint $p \in X$. We define the *contracting boundary* of a CAT(0) space X to be the subset of the visual boundary consisting of all contracting geodesics:

$$\partial_c X_e = \{\alpha(\infty) \in \partial X \mid \alpha \text{ is contracting with basepoint } e\}.$$

In order to topologise the contracting boundary we consider a collection of increasing subsets of the boundary,

$$\partial_c^n X_e = \{\gamma(\infty) \in \partial X \mid \gamma(0) = e, \gamma \text{ is a } n\text{-contracting ray}\},$$

one for each $n \in \mathbb{N}$. We topologise each $\partial_c^n X_e$ with the subspace topology from the visual boundary of X. We note that there is an obvious continuous inclusion map $i\colon \partial_c^m X_e \hookrightarrow \partial_c^n X_e$ for all $m < n$. We can topologise the whole boundary by taking the direct limit over these subspaces. Thus $\partial_c X_e = \varinjlim \partial_c^n X_e$ with the direct limit topology. Recall that this means a set U is open (resp. closed) in $\partial_c X_e$ if and only if $U \cap \partial_c^n X_e$ is open (resp. closed) for all $n \in \mathbb{N}$.

One nice property of this boundary is that if we fix a contracting constant n, then $\partial_c^n X_e$ is compact and behaves very much like the boundary of a hyperbolic space, an analogy we will build on in Section 4.3.

Another nice property is that the direct limit topology on $\partial_c X_e$ does not depend on the choice of basepoint so we will freely denote the contracting boundary as $\partial_c X$ without mention of basepoint.

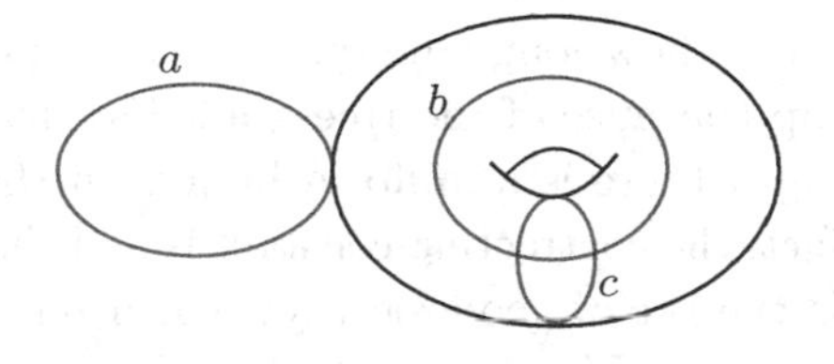

Figure 4.2 presentation complex of $\mathbb{Z} * \mathbb{Z}^2$

Examples

We will focus on examples of the contracting boundary, for a couple reasons. First, other more general incarnations of the Morse boundary, which we will see later, are homeomorphic to the contracting boundary in the case of a CAT(0) space. Second, there is a combinatorial characterization of contracting geodesics in CAT(0) cube complexes (see [CS2, Theorem 4.2]) that makes the computations more transparent.

Example 4.2.4 ($\mathbb{R}^n$, hyperbolic spaces) Revisiting the conversation about which spaces have contracting geodesics, when $X = \mathbb{R}^n$ for $n \geq 2$ or more generally the product of two unbounded CAT(0) spaces will have no contracting geodesics because any geodesic will be contained in a flat. This means that $\partial_c X$ will be empty. On the other hand, if X is CAT(0) and hyperbolic, then $\partial_c X$ will be the Gromov boundary because every ray will uniformly contracting. For example, $\partial\mathbb{H}^n$ will be homeomorphic to $\mathbb{S}^{n-1}$.

Example 4.2.5 (hyperbolic with a dash of flat) Now consider the space X formed by gluing a Euclidean half-plane to a bi-infinite geodesic γ in the hyperbolic plane $\mathbb{H}^2$. Picking $\gamma(0)$ to be our basepoint, we see that any geodesic in the half-plane (including γ) will not be Morse, but all the other rays in $\mathbb{H}^2$ will still be contracting. So the boundary will be the union of two open intervals. We can already see here how the contracting boundary differs from the Gromov boundary; it is not necessarily compact. In fact, Murray shows that the contracting boundary of a CAT(0) space is compact if any only if X is hyperbolic [Mu].

Example 4.2.6 ($\mathbb{Z} * \mathbb{Z}^2$) For a slightly more complicated example consider the space Y formed by the wedge product of a loop and a torus and its fundamental group $\pi_1(Y) = \mathbb{Z} * \mathbb{Z}^2 = \langle a \rangle * \langle b, c \rangle$. See Figure 4.2. The universal cover of Y, $\tilde{Y}$, is a CAT(0) cube complex [ChD]. Consider a geodesic ray γ in X. It is not hard to see that the contracting constant of γ only depends on how long subsegments of γ spend in the cosets of the $\mathbb{Z}^2$ subgroup (because of the tree-graded structure of $\tilde{Y}$), and it is only contracting if there is a uniform bound on the length of these subsegments. In fact, the contracting constant is half that uniform bound; thus each $\partial_c^n \tilde{Y}$ is the set of geodesic rays which have subsegments of length at most $2n$ in the $\mathbb{Z}^2$ cosets with the subspace topology. So $\partial_c \tilde{Y}$ is the direct limit over these spaces. We will see later, in Theorem 4.3.17, that the subspace of $\tilde{Y}$ formed by the union of all n-contracting geodesics with basepoint e is quasi-isometric to a proper simplicial tree and thus has boundary homeomorphic to a Cantor set.

Remark 4.2.7 (the contracting boundary is not in general first-countable) The topology of the contracting boundary is in general quite fine relative to the subspace topology. In fact, in [Mu] Murray shows that the space $\tilde{Y}$ as defined above is not first-countable (and thus not metrisable). We present Murray's argument here: choose the basepoint of $\tilde{Y}$ to be some lift of the wedge point in Y. We first note that the geodesic ray $\alpha = aaaa\ldots$ is 0-contracting. We define a collection of geodesic rays $\beta_i^j = a^i b^{2j} aaaa\ldots$. Note that the each of the β_i^j are j-contracting and not j'-contracting for any $j' < j$. If we fix a j, it is clear that the $\{\beta_i^j\}$ converge to α in the contracting boundary. Consider a new sequence $\{\beta_{f(j)}^j\}$, where f is any function $f\colon \mathbb{N} \to \mathbb{N}$. We note that the intersection of $\{\beta_{f(j)}^j\}$ with $\partial_c^n \tilde{Y}$ is finite for each $n \in \mathbb{N}$ and thus closed in the subspace topology and therefore in $\partial_c X$. Ergo, the $\{\beta_{f(j)}^j\}$ cannot converge to α in the contracting boundary.

A general fact for all first-countable spaces is that if you have a countable collection of sequences that all converge to the same point, it is always possible to pick a 'diagonal' sequence that also converges to that point. That is, if $\partial_c \tilde{Y}$ were first-countable, since $\{\beta_i^j\}$ converges to α for fixed j, then there would be a function $f\colon \mathbb{N} \to \mathbb{N}$ so that $\{\beta_{f(j)}^j\}$ converges to α. As we saw above, this cannot happen. Since all metric spaces are first countable, this means that the contracting boundary is not, in general, metrisable! In fact, in the same paper, Murray shows that if X is a CAT(0) space with a geometric action, then $\partial_c X$ is metrisable if any only if X is hyperbolic.

Example 4.2.8 (Croke–Kleiner) Recall that Croke and Kleiner produced an example of a right-angled Artin group that acts geometrically on two quasi-isometric CAT(0) spaces that have non-homeomorphic

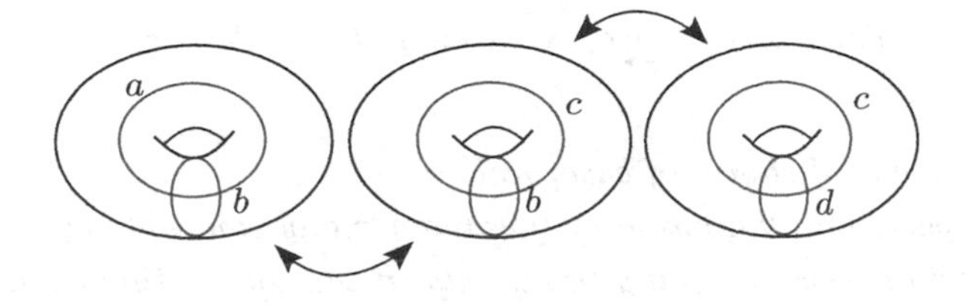

Figure 4.3 the Croke–Kleiner example

visual boundaries [CK]. Their example is the right-angled Artin group

$$A_\Gamma = \langle a, b, c, d \,|\, [a, b] = [b, c] = [c, d] = 1 \rangle .$$

The Salvetti complex of this group, S_Γ, is three tori with the middle torus glued to the other two along orthogonal curves corresponding to the generators b and c. See Figure 4.3. As in the example above, it follows from [ChD] that the universal cover of S_Γ, $\tilde{S}_\Gamma$, is a CAT(0) cube complex.

Let B_1 be the union of the (a, b)-torus and the (b, c)-torus in S_Γ and let $\tilde{B}_1$ be its inverse image in $\tilde{S}_\Gamma$. Each component of the inverse image decomposes as the direct product of the Cayley graph of the free group on two elements and $\mathbb{R}^2$. Thus the contracting boundary of each component of $\tilde{B}_1$ is empty. The same fact holds for B_2, the union of the (b, c)-torus and the (c, d)-torus in S_Γ. Croke and Kleiner refer to the components as "blocks". As in the example above, it follows that in order for a geodesic ray γ in $\tilde{S}_\Gamma$ to be contracting, there must be a uniform bound on the length of the subsegments of γ that intersect the blocks. The converse of this holds by Charney and Sultan's combinatorial characterization of contracting geodesics in CAT(0) cube complexes [CS2]. Thus, γ is contracting if and only if there is a uniform bound on the length of subsegments of γ that intersect a single block. We will revisit this in Example 4.2.10 after we describe some of the properties of the contracting boundary.

Properties of the boundary

This boundary has many desirable properties. First and foremost it is a quasi-isometry invariant. That is, if X and Y are complete CAT(0) spaces and $q\colon X \to Y$ is a quasi-isometry, then q induces a homeomorphism $\partial_q\colon \partial_c X \to \partial_c Y$. Furthermore since each $\partial_c^n X$ is compact, the boundary is σ-compact. Finally it is a visibility space. In summary:

Theorem 4.2.9 ([CS2]) *Given a complete* CAT(0) *space X, the con-*

tracting boundary, $\partial_c X = \varinjlim \partial_c^n X_e$, *equipped with the direct limit topology, is*

1. *independent of choice of basepoint;*
2. σ*-compact, i.e., the union of countably many compact sets;*
3. *a visibility space, i.e., any two points in the contracting boundary can be joined by a bi-infinite contracting geodesic; and*
4. *a quasi-isometry invariant.*

Example 4.2.10 (Croke–Kleiner redux) Croke and Kleiner produce the two spaces on which A_Γ acts geometrically by modifying the metric on $\tilde{S}_\Gamma$ in a very simple way: they skew the angles between the b and c curves making the (b, c)-cubes parallelograms. This is enough to change the homeomorphism type of the visual boundary. In fact, Wilson shows that any two distinct angles between the b and c curves produces non-homeomorphic boundaries [Wi]. Qing showed that if you keep the angles $\frac{\pi}{2}$ and change the side lengths of the cubes then the identity map does not induce a homeomorphism on the boundary [Q].

In the examples of Croke–Kleiner and Wilson, the parts of the boundary that change the homeomorphism type are the parts that come from the intersection of the blocks. These points do not appear in the contracting boundary. Neither do the points in the Qing example. The parts of the boundary that change are the rays that stay for longer and longer times in successive blocks. These examples suggest that the restriction to contracting rays may be optimal if you want a quasi-isometry invariant.

In [CS2], Charney and Sultan remarked that a quasi-isometry induces a bijection on the contracting rays and the set of contracting rays could be topologised with just the subspace topology from the visual boundary. They asked if a quasi-isometry would induce a homeomorphism on the boundary with this topology. In [Ca] Cashen answers this question in the negative:

Theorem 4.2.11 ([Ca]) *In general a quasi-isometry will not induce a homeomorphism on the contracting boundary with the subspace topology.*

Cashen's examples, though, are pathological in nature and open questions remain: if X, Y are CAT(0) spaces with cocompact isometry groups and if $\phi\colon X \to Y$ is a quasi-isometry, does ϕ induce a homeomorphism on the contracting boundary with the subspace topology? Are these two boundaries abstractly homeomorphic?

One can also ask when a homeomorphism between the boundaries of two spaces is induced by a quasi-isometry. In the case of hyperbolic spaces, Paulin [Pau] gives conditions under which this holds. Recent work of Charney, Cordes and Murray [CMC] gives some analogous conditions for the Morse boundary of a proper geodesic space.

Dynamics on the contracting boundary

In the realm of hyperbolic groups there is a classical notion of North–South dynamics:

Theorem 4.2.12 *If G is a hyperbolic group acting geometrically on a proper geodesic metric space X and if g is an element of infinite order, then for all open sets U and V with $g^{\infty} \in U$ and $g^{-\infty} \in V$ there exists an $n \in \mathbb{N}$ so that $g^n V^c \subset U$.*

It is well known that the classical notion of North–South dynamics of axial isometries on the visual boundary of a CAT(0) space fails because there are isometries that fix whole flats. Rank 1 isometries, though, do act on the visual boundary with North–South dynamics [Ba, Ham2]. Since rank 1 isometries of CAT(0) spaces are contracting, one might hope that North–South dynamics hold on the contracting boundary. This is not the case. Again we will present an example from [Mu].

Example 4.2.13 (Example 4.2.6 revisited) Again, we consider $\mathbb{Z} * \mathbb{Z}^2 = \langle a \rangle * \langle b, c \rangle$ and the spaces Y and $\tilde{Y}$. Let α be an axis for a. Let β_i be the geodesic defined by the word $a^{-i}b^i aaaa\ldots$. Again we note that the β_i do not converge to α in the contracting boundary because $\{\beta_i\} \cap \partial_c^D \tilde{Y}$ is finite for every D and thus $\{\beta_i\}$ is closed in $\partial_c \tilde{Y}$. We note that the set $V = (U(\alpha(-\infty), r, \epsilon) \cap \partial_c \tilde{Y}) \setminus \{\beta_i\}$ is an open set around $\alpha(-\infty)$ but for all $n \in \mathbb{N}$ we have $a^n \beta_n(t) \notin U(\alpha(\infty), r', \epsilon')$ for all $\epsilon' < r'$.

In [Mu] Murray does prove a weaker type of North–South dynamics on the contracting boundary.

Theorem 4.2.14 (Corollary 4.3 [Mu]) *Let X be a proper* CAT(0) *space and let G be a group acting geometrically on X. If g is a rank 1 isometry in G, i.e., a contracting element, U is an open neighbourhood of g^{∞} and K is a compact set in $\partial_c X \setminus g^{-\infty}$, then for sufficiently large n, $g^n(K) \subset U$.*

Murray also uses dynamical methods to prove a classical result that is known for the action of a hyperbolic group on its boundary.

Theorem 4.2.15 (Theorem 4.1 [Mu]) *Let G be a group acting geometrically on a proper* CAT(0) *space. Then either G is virtually $\mathbb{Z}$ or the G orbit of every point in the contracting boundary is dense.*

In Section 4.2.2 we will introduce a generalization of the contracting boundary to any proper geodesic space. It is unknown if any of these dynamical results hold in the more general settings.

4.2.2 Morse boundary

The Morse boundary, introduced by Cordes in [Co], generalizes the contracting boundary to the setting of proper geodesic spaces. This boundary retains many of the nice properties of the contracting boundary including quasi-isometry invariance and visibility. In the case of a proper CAT(0) space it is the contracting boundary, and in the case of a proper hyperbolic space it is the Gromov boundary. The generality in which this boundary is defined means it is a quasi-isometry invariant for every finitely generated group.

We will see later in Section 4.3 that there is a more general definition of the Morse boundary of a not-necessarily-proper geodesic space. This definition will use the Gromov product and will carry more structure, however it is often helpful to reduce to the case when the boundary can be defined by geodesics instead of sequences of points (as we will see in Section 4.4.1) for proper geodesic spaces.

Let X be a proper geodesic space and fix a basepoint $e \in X$. The *Morse boundary* of X, $\partial_M X$, is the set of all Morse geodesic rays in X (with basepoint e) up to asymptotic equivalence. To topologise the boundary, first fix a Morse gauge N and consider the subset of the Morse boundary that consists of all rays in X with Morse gauge at most N:

$$\partial_M^N X_e = \{\alpha(\infty) \mid \exists \beta \in \alpha(\infty) \text{ an } N\text{–Morse geodesic ray with } \beta(0) = e\}.$$

Unlike in the case of a CAT(0) space, the visual topology of the boundary of a proper geodesic space may not even be well-defined. So, instead, we take a page from the definition of the Gromov topology on the boundary of a hyperbolic space. This first step is a lemma in [Co] that says that two N-Morse geodesic rays with the same basepoint that fellow travel stay uniformly close and that the uniform bound, δ_N, only depends on N. Thus we can topologise this set in a similar manner as one does for the Gromov boundary of hyperbolic spaces: the topology is defined by a system of neighbourhoods, $\{V_n(\alpha) \mid n \in \mathbb{N}\}$, at a point α in $\partial_M^N X_e$. The sets $V_n(\alpha)$ are defined to be the set of geodesic rays γ with basepoint e and $d(\alpha(t), \gamma(t)) < \delta_N$ for all $t < n$. That is, two N-Morse rays are close in $\partial_M^N X_e$ if they stay closer than δ_N for a long time.

Let $\mathcal{M}$ be the set of all Morse gauges. We put a partial ordering on $\mathcal{M}$ so that for two Morse gauges $N, N' \in \mathcal{M}$, we say $N \leq N'$ if and only if $N(\lambda, \epsilon) \leq N'(\lambda, \epsilon)$ for all $\lambda, \epsilon \in \mathbb{N}$. We define the Morse boundary of X to be

$$\partial_M X_e = \varinjlim_{\mathcal{M}} \partial_M^N X_e$$

with the induced direct limit topology, i.e., a set U is open in $\partial_M X_e$ if and only if $U \cap \partial_M^N X_e$ is open for all N.

The Morse boundary retains almost all of the properties of the contracting boundary. The one exception is that it is open whether or not the Morse boundary is σ-compact, because the direct limit is *a priori* over an uncountable set.

Theorem 4.2.16 ([Co]) *Given a proper geodesic space X, the Morse boundary, $\partial_M X = \varinjlim \partial_M^N X_e$, equipped with the direct limit topology, is*

1. *a visibility space, i.e., any two points in the Morse boundary can be joined by a bi-infinite Morse geodesic;*
2. *independent of choice of basepoint;*
3. *a quasi-isometry invariant; and*
4. *homeomorphic to the Gromov boundary if X is hyperbolic and to the contracting boundary if X is* CAT(0).

One useful property of the Morse boundary is that compact subsets consist of uniformly Morse geodesics [Mu, CoD]. Furthermore, as in the case of the contracting boundary, a group has a compact Morse boundary if and only if it is hyperbolic [CoD].

4.3 (Metric) Morse boundary and stability

An alternative approach to understanding "hyperbolic directions" in a metric space is to understand "hyperbolic" or quasi-convex subgroups/subspaces. In the case of hyperbolic groups, quasi-convex subgroups are finitely generated and undistorted. Furthermore these properties are preserved under quasi-isometry. In a general group, though, quasi-convexity depends on a choice of generating set and is not preserved by quasi-isometry. Thus in an effort to preserve these qualities, we look at a stronger notion of quasi-convexity:

Definition 4.3.1 We say a quasi-convex subspace Y of a geodesic metric space X is *N-stable* if every pair of points in Y can be connected by a geodesic that is N-Morse in X. We say that a subgroup is stable if it is stable as a subspace.

Remark 4.3.2 (Relationship with Durham–Taylor definition) It is important to note that this is a generalization of the original definition of stability given by Durham–Taylor in [DT]. The definition above detects the same collection of stable subsets up to quasi-isometry, and the two definitions coincide for subgroups of finitely generated groups [CH2, Lemma 3.8].

Durham–Taylor prove that the collection of stable subgroups of mapping class groups are precisely those that are convex-cocompact in the sense of Farb–Mosher [FM, DT]. These subgroups are well studied and have important connections to the geometry of Teichmüller space, the curve complex and surface group extensions. In the setting of right-angled Artin groups, Koberda, Manghas, and Taylor classify all the stable subgroups. They prove that these subgroups are all free [KMT]. For more on stable subgroups, see Section 4.4.

We will see in this section that the notions of the Morse boundary and of stability can be united. We will do this by viewing any geodesic metric space as the union of stable subsets that are indexed by Morse gauges N and are hyperbolic (with hyperbolicity constant depending only on N). We will define these subspaces in Section 4.3.2, but before this we will recall some definitions.

4.3.1 Sequential boundary, capacity dimension & asymptotic dimension

Definition 4.3.3 Let X be a metric space and let $x, y, z \in X$. The *Gromov product* of x and y with respect to z is defined as

$$(x \cdot y)_z := \frac{1}{2}\left(d(z,x) + d(z,y) - d(x,y)\right).$$

Let (x_n) be a sequence in X. We say (x_n) converges at infinity if $(x_i \cdot x_j)_e \to \infty$ as $i, j \to \infty$. Two convergent sequences $(x_n), (y_m)$ are said to be *equivalent* if $(x_i \cdot y_j) \to \infty$ as $i, j \to \infty$. We denote the equivalence class of (x_n) by $\lim x_n$.

The *sequential boundary* of X, denoted ∂X, is defined to be the set of convergent sequences considered up to equivalence.

Definition 4.3.4 (4-point definition of hyperbolicity; [BH], Definition 1.20) Let X be a (not necessarily geodesic) metric space. We say X is *δ–hyperbolic* if for all w, x, y, z we have

$$(x \cdot y)_w \geq \min\{(x \cdot z)_w, (z \cdot y)_w\} - \delta.$$

If X is δ–hyperbolic, we may extend the Gromov product to ∂X in the following way:

$$(x \cdot y)_e = \sup\left(\liminf_{m,n\to\infty} \{(x_n \cdot y_m)_e\}\right),$$

where $x, y \in \partial X$ and the supremum is taken over all sequences (x_i) and (y_j) in X such that $x = \lim x_i$ and $y = \lim y_j$.

Recall that a metric d on ∂X is said to be *visual* (with parameter $\varepsilon > 0$) if there exist $k_1, k_2 > 0$ such that $k_1 \exp(-\varepsilon(x \cdot y)_e) \leq d(x, y) \leq k_2 \exp(-\varepsilon(x \cdot y)_e)$, for all $x, y \in \partial X$.

Let $x, y \in \partial X$. As a shorthand we define $\rho_\epsilon(x, y) := \exp\left(-\varepsilon(x \cdot y)_e\right)$.

Theorem 4.3.5 *[GdlH, Section 7.3] Let X be a δ–hyperbolic space. If $\varepsilon' = \exp(2\delta\varepsilon) - 1 \leq \sqrt{2} - 1$, then we can construct a visual metric d on ∂X such that*

$$(1 - 2\varepsilon')\rho_\varepsilon(x, x') \leq d(x, x') \leq \rho_\varepsilon(x, x').$$

Visual metrics on a hyperbolic space are all quasi-symmetric. A quasi-symmetry is a map that is a generalization of bi-Lipschitz map: instead of controlling how much the diameter of a set can change, a quasi-symmetry preserves only the relative sizes of sets.

Definition 4.3.6 A homeomorphism $f\colon (X, d) \to (Y, d')$ is said to be *quasi-symmetric* if there exists a homeomorphism $\eta : \mathbb{R} \to \mathbb{R}$ such that for all distinct $x, y, z \in X$,

$$\frac{d'(f(x), f(y))}{d'(f(x), f(z))} \leq \eta\left(\frac{d(x, y)}{d(x, z)}\right).$$

One natural quasi-isometry invariant assigned to the boundary of a hyperbolic space is the *capacity dimension*. This invariant was introduced by Buyalo in [Bu] and it is sometimes known as linearly-controlled dimension.

Let $\mathcal{U}$ be an open covering of a metric space X. Given $x \in X$, we let

$$L(\mathcal{U}, x) = \sup\{d(x, X\backslash U) \mid U \in \mathcal{U}\}$$

be the *Lebesgue number of* $\mathcal{U}$ *at* x and $L(\mathcal{U}) = \inf_{x \in X} L(\mathcal{U}, x)$ the Lebesgue number of $\mathcal{U}$. The *multiplicity of* $\mathcal{U}$, $m(\mathcal{U})$, is the maximal number of members of $\mathcal{U}$ with non-empty intersection.

Definition 4.3.7 ([BS]) The *capacity dimension*, $\mathrm{cdim}(X)$, of a metric space X is the minimal integer m with the following property:

There exists some $\delta \in (0,1)$ such that for every sufficiently small $r > 0$ there is an open covering $\mathcal{U}$ of X by sets of diameter at most r with $L(\mathcal{U}) \geq \delta r$ and $m(\mathcal{U}) \leq m + 1$.

The capacity dimension is similar to the covering dimension in that it is an infimum over open covers, but the capacity dimension necessitates metric information: given an open cover $\mathcal{U}$ the capacity dimension requires a linear relationship between the $\sup_{U \in \mathcal{U}}\{\mathrm{diam}(U)\}$ and $L(\mathcal{U})$. For more information see [BS].

In [Bu, Corollary 4.2], Buyalo shows that the capacity dimension of a metric space is a quasi-symmetry invariant and since quasi-isometries of hyperbolic spaces induce quasi-symmetries on the boundaries, this shows that the capacity dimension of the boundary is an invariant of a hyperbolic space.

Another quasi-isometry-invariant notion of dimension one can assign to a metric space is the asymptotic dimension. This notion was introduced by Gromov in [Grom1] and is a coarse version of the topological dimension.

Definition 4.3.8 A metric space X has *asymptotic dimension at most* n ($asdim(X) \leq n$), if for every $R > 0$ there exists a cover of X by uniformly bounded sets such that every metric R–ball in X intersects at most $n + 1$ elements of the cover. We say X has *asymptotic dimension* n if $asdim(X) \leq n$ but $asdim(X) \nleq n - 1$.

A celebrated theorem of Yu shows that groups with finite asymptotic dimension satisfy both the coarse Baum–Connes and the Novikov conjectures [Y]. Many classes of groups have been shown to have finite asymptotic dimension, including: hyperbolic groups [R], relatively hyperbolic groups whose parabolic subgroups are of finite dimension [Osi1], mapping class groups [BBF], and cubulated groups [Wr]. But exact bounds are often hard to calculate.

Buyalo and Lebedeva used the capacity dimension to prove a conjecture of Gromov: the asymptotic dimension of any hyperbolic group is the topological dimension of its boundary plus one [BL].

4.3.2 Stable strata

Definition 4.3.9 (stable stratum) Let X be a geodesic metric space and let $e \in X$. We define $X_e^{(N)}$ to be the set of all points in X that can be joined to e by an N-Morse geodesic.

What do these stable strata look like? First, we can easily see that given $x \in X$ there exists N such that $x \in X_e^{(N)}$ — by choosing $N = N(\lambda, \epsilon) = \lambda d(e, x) + \epsilon$. So the collection of $X_e^{(N)}$ covers X. If X is hyperbolic, since all geodesics are N-Morse for some N, we have that $X = X_e^{(N)}$ for that N. On the other hand, if X is a space with no infinite Morse geodesics (e.g., groups satisfying non-trivial laws or products of unbounded spaces) then the $X_e^{(N)}$ are just sets of bounded diameter. For groups with mixed geometries this is a hard question to answer. We will see later in the survey (Sections 4.3.4 and 4.3.5) that (up to quasi-isometry) we can begin to understand these subspaces for some groups.

One thing we do know is that these subspaces are hyperbolic. A standard argument shows that if $x, y \in X_e^{(N)}$ then the geodesic triangle in X formed by x, y, e is $4N(3,0)$-slim [Co, Lemma 2.2], proving these spaces are hyperbolic. Note, though, that the geodesic $[x, y]$ is not necessarily contained in $X_e^{(N)}$. So since the $X_e^{(N)}$ are not necessarily geodesic, we use the 4-point definition of hyperbolicity and conclude they are hyperbolic.

Using the fact that these triangles are slim, one can also show that if $x, y \in X_e^{(N)}$, then $[x, y]$ is N'-Morse, where N' depends only on N [Co, Lemma 2.3]. This shows that the $X_e^{(N)}$ are not only quasi-convex, but they are N'-stable!

Furthermore, as each $X_e^{(N)}$ is hyperbolic, we may consider its Gromov boundary $\partial_s X_e^{(N)}$, and the associated visual metric $d_{(N)}$. (See [BS, Section 2.2] for a careful treatment of the sequential boundary of a hyperbolic space that is not necessarily geodesic.) This metric is unique up to quasi-symmetry.

Natural maps between strata have nice properties: the natural inclusion $X_e^{(N)} \subseteq X_e^{(N')}$ induces a map $\partial_s X_e^{(N)} \to \partial_s X_e^{(N')}$ which is a quasi-symmetry onto its image. Additionally, if we have a quasi-isometry $q \colon X \to Y$, then for every N there exists an N' such that $q(X_e^{(N)}) \subseteq Y_{q(e)}^{(N')}$. This induces an embedding $\partial q \colon \partial_s X_e^{(N)} \to \partial_s Y_{q(e)}^{(N')}$ which is a quasi-symmetry onto its image. We will see in Section 4.3.3 that these properties will be useful in defining new quasi-isometry invariants.

Finally, if X is also a proper metric space, then for each N there is a homeomorphism between $\partial_M^N X$ (as defined in Section 4.2.2 with geodesics) and $\partial_s X_e^{(N)}$, thus the Morse boundary is homeomorphic to direct limit over $\partial_s X_e^{(N)}$ as topological spaces.

In summary:

Theorem 4.3.10 ([CH2]) *Let X, Y be geodesic metric spaces and let $e \in X$. The family of subsets $X_e^{(N)}$ of X indexed by functions $N: \mathbb{R}_{\geq 1} \times \mathbb{R}_{\geq 0} \to \mathbb{R}_{\geq 0}$ enjoys the following properties:*

1 *(covering) $X = \bigcup_N X_e^{(N)}$.*

2 *(partial order) If $N \leq N'$, then $X_e^{(N)} \subseteq X_e^{(N')}$.*

3 *(hyperbolicity) Each $X_e^{(N)}$ is hyperbolic in the sense of Definition 4.3.4.*

4 *(stability) Each $X_e^{(N)}$ is N'-stable, where N' depends only on N.*

5 *(universality) Every stable subset of X is a quasi-convex subset of some $X_e^{(N)}$.*

6 *(boundary) The sequential boundary $\partial_s X_e^{(N)}$ can be equipped with a visual metric which is unique up to quasi-symmetry. An inclusion $X_e^{(N)} \subseteq X_e^{(N')}$ induces a map $\partial_s X_e^{(N)} \to \partial_s X_e^{(N')}$ which is a quasi-symmetry onto its image.*

7 *(generalizing the Gromov boundary) If X is hyperbolic, then $X = X_e^{(N)}$ for all N sufficiently large, and $\partial_s X_e^{(N)}$ is quasi-symmetric to the Gromov boundary of X.*

8 *(generalizing the Morse boundary) If X is proper, then its Morse boundary is equal to the direct limit of the $\partial_s X_e^{(N)}$ as topological spaces.*

9 *(behaviour under quasi-isometry) If $q: X \to Y$ is a quasi-isometry, then for every N there exists an N' such that $q(X_e^{(N)}) \subseteq Y_{q(e)}^{(N')}$ and there is an induced embedding $\partial q: \partial_s X_e^{(N)} \to \partial_s Y_{q(e)}^{(N')}$ that is a quasi-symmetry onto its image.*

There is a stronger version of 4.3.10.9, but we first require some additional terminology.

Let X, Y be geodesic metric spaces, and let $x \in X$ and $y \in Y$. We say X is *stably subsumed* by Y (denoted $X \hookrightarrow_s Y$) if for every N there exists a quasi-isometric embedding $X_x^{(N)} \to Y_y^{(N')}$ for some N'. By 4.3.10.9, this property is independent of the choice of x, y. We say X and Y are *stably equivalent* (denoted $X \sim_s Y$) if they stably subsume each other. It is easy to see that two spaces are stably equivalent if and only if they have the same collection of stable subsets up to quasi-isometry.

Given a geodesic metric space X, we consider the collection of boundaries $\left(\partial_s X_e^{(N)}\right)$ equipped with visual metrics as the *metric Morse boundary* of X.

We say that one collection of spaces $(A_i)_{i\in I}$ is *quasi-symmetrically subsumed* by another $(B_j)_{j\in J}$ (denoted $(A_i) \hookrightarrow_{qs} (B_j)$) if, for every i there exists a j and an embedding $A_i \to B_j$ that is a quasi–symmetry onto its image. Two collections are *quasi-symmetrically equivalent* (denoted $(A_i) \sim_{qs} (B_j)$) if $(A_i) \hookrightarrow_{qs} (B_j)$ and $(B_j) \hookrightarrow_{qs} (A_i)$).

Theorem 4.3.10.9' *Let X, Y be geodesic metric spaces, let $x \in X$ and $y \in Y$. Then $X \hookrightarrow_s Y$ if and only if $\left(\partial X_x^{(N)}\right) \hookrightarrow_{qs} \left(\partial Y_y^{(N)}\right)$.*

Corollary 4.3.11 *Quasi-isometric geodesic metric spaces are stably equivalent and have quasi-symmetrically equivalent metric Morse boundaries. In particular, the metric Morse boundary is invariant under change of basepoint.*

Stable equivalence is a much weaker notion than quasi-isometry. All spaces with no infinite Morse geodesic rays will be stably equivalent to a point! On the other hand, Cordes–Hume show that the mapping class group and Teichmüller space with the Teichmüller metric are stably equivalent (Theorem 4.3.16), while it is well known that they are not quasi-isometric.

4.3.3 Stable equivalence invariants

We can now define two stable equivalence invariants using the notions discussed in Section 4.3.1; in particular, the capacity dimension and the asymptotic dimension. The definitions will rely on Theorem 4.3.10.

Definition 4.3.12 (stable asymptotic dimension) The *stable asymptotic dimension* of X ($\operatorname{asdim}_s(X)$) is the supremal asymptotic dimension of a stable subset of X.

We can see that by universality it is possible to consider only the maximal asymptotic dimension of the $X_e^{(N)}$. One obvious but useful bound is that the stable asymptotic dimension is bounded from above by the asymptotic dimension.

Definition 4.3.13 (Morse capacity dimension) The *Morse capacity dimension* of X ($\operatorname{cdim}_{\partial_M}(X)$) is the supremal capacity dimension of spaces in the metric Morse boundary. We say that the empty set has capacity dimension -1.

By Theorem 4.3.10.6 we know that the inclusion map $X_e^{(N)} \hookrightarrow X_e^{(N')}$ induces a map $\partial X_e^{(N)} \hookrightarrow \partial X_e^{(N')}$ that is a quasi-symmetry onto its image. So the Morse capacity dimension is well-defined.

It follows from Theorem 4.3.10 that these notions are invariant under change of basepoint and are stable equivalence invariants and thus quasi-isometry invariants. It also follows that the stable dimension of a hyperbolic space is precisely its asymptotic dimension and the Morse capacity dimension of a hyperbolic space is the capacity dimension of its boundary equipped with some visual metric.

Remark 4.3.14 (conformal dimension) Since the conformal dimension (introduced by Pansu in [Pan]) is also a quasi-symmetry invariant, one could also define a conformal dimension of the Morse boundary in the same manner and with the properties listed above. However it is often harder to compute.

By using bounds bounds proved in the hyperbolic setting, we get a nice relationship between these two dimensions ([Bu, Theorem 1.1], [MS, Proposition 3.6]).

Proposition 4.3.15 *Let X be a geodesic metric space. Then*

$$asdim_s(X) - 1 \leq cdim_{\partial_M}(X) \leq asdim_s(X).$$

4.3.4 Calculations in finitely generated groups

We now present some calculations in finitely generated groups wherein Morse geodesics have been characterized in some way: mapping class groups, right-angled Artin groups, and $C'(1/6)$ graphical small cancellation groups. As we will see, a recurring way to calculate an upper bound on the stable asymptotic dimension is to show that each stable stratum embeds into a hyperbolic space that is better understood.

Mapping class groups and Teichmüller space

Assume that Σ is an orientable surface of finite type. Let $\mathcal{T}(\Sigma)$ be the Teichmüller space of Σ with the Teichmüller metric. The following result is a generalization of a result in [Co] which shows that $\partial_M \mathrm{MCG}(\Sigma)$ is homeomorphic to $\partial_M \mathcal{T}(\Sigma)$.

Theorem 4.3.16 ([CH2]) $\mathrm{MCG}(\Sigma)$ *and* $\mathcal{T}(\Sigma)$ *are stably equivalent, thus* $asdim_s(\mathrm{MCG}(\Sigma)) = asdim_s(\mathcal{T}(\Sigma))$. *Furthermore,* $asdim_s(\mathrm{MCG}(\Sigma))$ *and* $cdim_{\partial_M}\mathrm{MCG}(\Sigma)$ *are bounded above linearly in the complexity of* Σ.

An upper bound on the stable asymptotic dimension of mapping class groups can be obtained via the bounds on the asymptotic dimension for mapping class groups obtained by Bestvina–Bromberg–Fujiwara in [BBF] or by Behrstock–Hagen–Sisto in [BHS], which are respectively exponential and quadratic in the complexity of the surface. To show there is a linear bound on the stable asymptotic dimension, Cordes–Hume show each stable subset of $\mathrm{MCG}(\Sigma)$ quasi-isometrically embeds into the curve graph. They then use the bound found by Bestvina–Bromberg on the asymptotic dimension of the curve graph [BeB].

Leininger and Schleimer prove that for every n there is a surface Σ such that the Teichmüller space $\mathcal{T}(\Sigma)$ contains a stable subset quasi–isometric to $\mathbb{H}^n$ [LS]. This fact gives a lower bound on the stable dimension, which is at best logarithmic in the complexity, but also shows that Teichmüller spaces can have arbitrarily high stable asymptotic dimension. Since $\mathcal{T}(\Sigma)$ is stably equivalent to $\mathrm{MCG}(\Sigma)$, we see that $\mathrm{MCG}(\Sigma)$ contains a stable subset quasi-isometric to $\mathbb{H}^n$. The only known explicit examples of convex cocompact subgroups of mapping class groups are virtually free groups [DKL, KLS, KMT, Min]. Although the results of Leininger–Schleimer do not provide any non-virtually-free convex cocompact subgroups, the fact that $\mathrm{asdim}_s(\mathrm{MCG}(\Sigma)) > 1$ for some surfaces shows that there is no purely geometric obstruction to the existence of a non-free convex cocompact subgroup of $\mathrm{MCG}(\Sigma)$.

Right-angled Artin groups

Koberda, Manghas and Taylor classify the stable subgroups of all right-angled Artin groups [KMT]. They show that these subgroups are always free. The next theorem is the natural analogue for stable subspaces.

Theorem 4.3.17 ([CH2]) *Let X be a Cayley graph of a right-angled Artin group. Then every stable subset of X is quasi-isometric to a proper tree. In particular, X is stably equivalent to a line if the group is abelian of rank* 1, *a point if it is abelian of rank* $\neq 1$ *and a regular trivalent tree otherwise.*

The proof of this theorem follows in a very similar manner to the proof of Theorem 4.3.16. In this case each stable subset of X embeds into the contact graph (defined by Hagen [Hag]), which is a quasi-tree. As a result, each $X_e^{(N)}$ is quasi-isometric to a proper tree. The proof is completed by calling on the universality condition in Theorem 4.3.10. By the universality of stable subsets (Theorem 4.3.10.5), we know that any stable subgroup of a right-angled Artin group is quasi-isometric to a proper simplicial tree. Thus, since groups that are quasi-isometric to trees are virtually free [GdlH, Corollary 7.19], they recover a result of [KMT]: stable subgroups of right–angled Artin groups are free.

Small cancellation

We move to the realm of graphical small cancellation groups. Graphical small cancellation theory is a generalization introduced by Gromov [Grom2] in order to construct groups whose Cayley graphs contain certain prescribed subgraphs, in particular one can construct "Gromov monster" groups, those with a Cayley graph that coarsely contains expanders [AD, Osa]. These monster groups cannot be coarsely embedded into a Hilbert space, and they are the only known counterexamples to the Baum–Connes conjecture with coefficients [HLS].

Theorem 4.3.18 ([CH2]) *Let X be the Cayley graph of a graphical $C'(1/6)$ small cancellation group. Then we have $asdim_s(X) \leq 2$ and $cdim_{\partial_M}(X) \leq 1$.*

Note that this is optimal, as fundamental groups of higher genus surfaces are hyperbolic with asymptotic dimension 2 and admit $C'(\frac{1}{6})$ graphical small cancellation presentations.

Again, we work with the stable strata. Each of the strata embeds quasi-isometrically into a finitely presented classical $C'(1/6)$ small cancellation group. These groups are hyperbolic with asymptotic dimension at most 2 and the capacity dimension of their Gromov boundaries is at most 1.

4.3.5 Relatively hyperbolic groups

Given $(G, \mathcal{H})$ a relatively hyperbolic group pair, one natural question one might ask is if there is a relationship between the Morse boundary of G and the Morse boundary of some peripheral subgroup $H \in \mathcal{H}$. Theorem 7.6 in [T], shows that there is a very nice relationship:

Theorem 4.3.19 ([T]) *Let $(G, \mathcal{H})$ be a finitely generated relatively hyperbolic group pair. Then for each $H \in \mathcal{H}$ the boundary map induced by the inclusion $\partial_M \iota \colon \partial_M H \to \partial_M G$ is a topological embedding.*

Relationship with Bowditch boundary

Let $(G, \mathcal{H})$ be a relatively hyperbolic group pair. Let Γ be the Cayley graph of G and let $\hat{\Gamma}$ be the coned-off Cayley graph. To each triangle with two vertices in Γ and one as a cone point, glue in a hyperbolic spike (the subset $[0,1] \times [1,\infty)$ in the upper half-space model of $\mathbb{H}^2$). This space is Gromov hyperbolic (if and only if $(G,\mathcal{H})$ is a relatively hyperbolic group pair). This construction is due to Bowditch [Bo3] and the boundary of this cusped space is called the Bowditch boundary, which we denote $\partial(G,\mathcal{H})$. We note that the cone points are points in the Bowditch boundary and we call these points *parabolic*. A point in the Bowditch boundary that is also a point in $\partial\hat{\Gamma}$ is called a *non-parabolic point*.

In [T], Tran defines a G-equivariant continuous map $f\colon \partial_M G \to \partial(G,\mathcal{H})$ and describes the relationship between points in the Morse boundary of G and parabolic and non-parabolic points in the Bowditch boundary of $(G,\mathcal{H})$.

Theorem 4.3.20 ([T]) *Let $(G,\mathcal{H})$ be a finitely generated relatively hyperbolic group pair. Then there is a G-equivariant continuous map $f\colon \partial_M G \to \partial(G,\mathcal{H})$ with the following properties:*

1. *f maps the set of non-peripheral limit points of $\partial_M G$ injectively into the set of non-parabolic points of $\partial(G,\mathcal{H})$;*
2. *f maps peripheral limit points of the same type in $\partial_M G$ to the same parabolic point in $\partial(G,\mathcal{H})$.*

In particular, if the Morse boundary of each peripheral subgroup is empty, then f maps $\partial_M G$ injectively into the set of non-parabolic points of $\partial(G,\mathcal{H})$.

Distinguishing quasi-isometry types

In [Osi1], Osin shows that relatively hyperbolic groups inherit finite asymptotic dimension from their maximal parabolic subgroups. This is also true for the stable asymptotic dimension:

Theorem 4.3.21 ([CH2]) *Let G be a finitely generated group that is hyperbolic relative to $\mathcal{H}$. Then $asdim_s(G) < \infty$ if and only if $asdim_s(H) < \infty$ for all $H \in \mathcal{H}$.*

In [CH1] Cordes–Hume focus on relatively hyperbolic groups. In this paper they suggest an approach to answering the following question, which appears in [BDM]: how may we distinguish non-quasi-isometric relatively hyperbolic groups with non-relatively hyperbolic peripheral subgroups when their peripheral subgroups are quasi-isometric?

Using small cancellation theory over free products, Cordes–Hume construct quasi-isometrically distinct one-ended relatively hyperbolic groups which are all hyperbolic relative to the same collection of groups. These groups are distinguished using a notion similar to stable asymptotic dimension; rather than $X_e^{(N)}$, we use stable subsets that "avoid" the left cosets of the peripheral subgroups.

Theorem 4.3.22 ([CH1]) *Let $\mathcal{H}$ be a finite collection of finitely generated groups each of which has finite stable dimension or is non-relatively hyperbolic. Then there is an infinite family of non-quasi-isometric 1-ended groups $(G_n)_{n\in\mathbb{N}}$, where each G_n is hyperbolic relative to $\mathcal{H}$.*

4.4 Stable subgroups

We will start our discourse on stable subgroups with some motivation from Kleinian groups and mapping class groups.

A non-elementary discrete (Kleinian) subgroup $\Gamma < \mathrm{PSL}_2(\mathbb{C})$ determines a minimal Γ-invariant closed subspace $\Lambda(\Gamma)$ of the Riemann sphere called its *limit set*, and taking the convex hull of $\Lambda(\Gamma)$ determines a convex subspace of $\mathbb{H}^3$ with a Γ-action. A Kleinian group Γ is called *convex cocompact* if it acts cocompactly on this convex hull or, equivalently, if any Γ-orbit in $\mathbb{H}^3$ is quasiconvex.

Originally defined by Farb–Mosher [FM] and later developed further by Kent–Leininger [KL2] and Hamenstädt [Ham1], a subgroup $H < \mathrm{MCG}(\Sigma)$ is called convex cocompact if and only if any H-orbit in $\mathcal{T}(\Sigma)$, the Teichmüller space of Σ with the Teichmüller metric, is quasiconvex, or if H acts cocompactly on the weak hull of its limit set $\Lambda(H) \subset \mathbb{P}\mathcal{MF}(\Sigma)$ in the Thurston compactification of $\mathcal{T}(\Sigma)$. This notion is important because convex cocompact subgroups $H < \mathrm{MCG}(\Sigma)$ are precisely those that determine Gromov hyperbolic surface group extensions. Furthermore, Farb–Mosher show that if there is a purely pseudo-Anosov subgroup of $\mathrm{MCG}(\Sigma)$ that is not convex cocompact, then this subgroup would be a counterexample to Gromov's conjecture that every group with a finite Eilenberg–Mac Lane space and no Baumslag–Solitar subgroups is hyperbolic (see [KL1] for more information).

In both of these examples, convex cocompactness is characterized equivalently by both a quasi-convexity condition and an asymptotic boundary condition. In [DT], Durham–Taylor introduced stability in order to characterize convex cocompactness in $\mathrm{MCG}(\Sigma)$ by a quasiconvexity condition intrinsic to the geometry of $\mathrm{MCG}(\Sigma)$, and this condition naturally generalizes the above quasi-convexity characterizations of convex cocompactness to any finitely generated group. There has been much recent work to characterize stable subgroups of important groups. The theorem below is a brief summary of this work.

Theorem 4.4.1 *Let the pair (G, H) of a finitely generated group G and a subgroup H satisfy one of the following:*

1 G is hyperbolic and H is quasi-convex;
2 G is relatively hyperbolic and H is finitely generated and embeds quasi-isometrically in the coned-off graph in the sense of [F];
3 $G = A(\Gamma)$ is a right-angled Artin group with Γ a finite graph that is not a join and H is a finitely generated subgroup quasi-isometrically embedded in the extension graph [KMT];
4 $G = \mathrm{MCG}(\Sigma)$ and H is a convex cocompact subgroup in the sense of [FM];
5 $G = \mathrm{Out}(F_n)$ for $n \geq 3$ and H is a convex cocompact subgroup in the sense of [HH];
6 H is hyperbolic and hyperbolically embedded in G.

Then H is stable in G. Moreover for (1), (3), and (4), the reverse implication holds.

Item (1) follows easily by checking the definition. Item (2) is due to [ADT], (3) due to [KMT], (4) is due to [DT], (5) is again [ADT], and (6) follows from [Si].

4.4.1 Boundary convex cocompactness

There is an alternative notion of stability to the Durham–Taylor characterization, using the Morse boundary to define an asymptotic property for subgroups of finitely generated groups called *boundary convex cocompactness* which generalizes the classical boundary characterization of convex cocompactness from Kleinian groups. This is presented in [CoD].

Let G be a finitely generated group acting by isometries on a proper geodesic metric space X. Fix a basepoint $e \in X$. In order to define this boundary characterization we first need to define a limit set.

Definition 4.4.2 ($\Lambda(G)$) The *limit set* of the G-action on $\partial_M X$ is

$$\Lambda_e(G) = \left\{\lambda \in \partial_M X \;\middle|\; \exists N \;\&\; (g_k) \subset G \text{ s.t. } (g_k \cdot e) \subset X_e^{(N)} \;\&\; \lim g_k \cdot e = \lambda\right\}$$

where the limit is taken in the hyperbolic space $X_e^{(N)}$.

That is, the limit set $\Lambda_e(G)$ is the set of points that can be represented by sequences of uniformly Morse G-orbit points; note that $\Lambda_e(G)$ is obviously G-invariant. It is also invariant under change of basepoint.

Given some $\lambda \in \Lambda_e(G)$ we know there is a sequence $(g_k \cdot e)$ so that $\lim g_k \cdot e = \lambda$. We can produce a geodesic from this sequence in a standard way: for each k let α_k be an N-Morse geodesic joining e and $g_k \cdot e$. It follows from Arzelà–Ascoli that there is a subsequence $(\alpha_{i(k)})$ that converges (uniformly on compact sets) to a geodesic ray that is N-Morse by [Co, Lemma 2.8]. We call such a ray γ_λ. This map defines a homeomorphism between the Morse boundary defined by sequences and the Morse boundary defined by geodesics [CH2, Theorem 3.14]. From this perspective, by Theorem 4.2.16 we know that there exists a bi-infinite Morse geodesic joining any two distinct points in the limit set. This is a starting point for defining the weak hull, but we want to define it by taking *all* geodesics with distinct endpoints in $\Lambda_e(G)$. Formalizing this motivates the next definition:

Definition 4.4.3 (asymptotic, bi-asymptotic) Let $\gamma\colon (-\infty, \infty) \to X$ be a bi-infinite geodesic in X with $\gamma(0)$ a closest point to e along γ. Let $\lambda \in \partial X_e^{(N)}$. We say γ is *forward asymptotic* to λ if for any N-Morse geodesic ray $\gamma_\lambda : [0, \infty) \to X$ with $\gamma_\lambda(0) = e$, there exists $K > 0$ such that

$$d_{Haus}(\gamma([0,\infty)), \gamma_\lambda([0,\infty))) < K.$$

We define *backwards asymptotic* similarly. If γ is forwards and backwards asymptotic to λ, λ', respectively, then we say γ is *bi-asymptotic* to (λ, λ').

Now for the definition of the weak hull.

Definition 4.4.4 (weak hull) The *weak hull* of G in X based at $e \in X$, denoted $\mathfrak{H}_e(G)$, is the collection of all bi-infinite rays γ that are bi-asymptotic to (λ, λ') for some $\lambda \neq \lambda' \in \Lambda_e(G)$.

An important fact about the weak hull is that if $|\Lambda_e(G)| \geq 2$, then $\mathfrak{H}_e(G)$ is nonempty and G-invariant.

We are now ready to define boundary convex cocompactness:

Definition 4.4.5 (boundary convex cocompactness) We say that G acts *boundary convex cocompactly* on X if the following conditions hold:

1 G acts properly on X;
2 for some (any) $e \in X$, $\Lambda_e(G)$ is nonempty and compact;
3 for some (any) $e \in X$, the action of G on $\mathfrak{H}_e(G)$ is cobounded.

Definition 4.4.6 (Boundary convex cocompactness for subgroups) Let G be a finitely generated group. We say $H < G$ is *boundary convex cocompact* if H acts boundary convex cocompactly on any Cayley graph of G with respect to a finite generating set.

Theorem 4.4.7 ([CoD]) *Let G be a finitely generated group. Then $H < G$ is boundary convex cocompact if and only if H is stable in G.*

Theorem 4.4.7 is an immediate consequence of the following stronger statement:

Theorem 4.4.8 ([CoD]) *Let G be a finitely generated group acting by isometries on a proper geodesic metric space X. Then the action of G is boundary convex cocompact if and only if some (any) orbit of G in X is a stable embedding.*

In both cases, G is hyperbolic and any orbit map $orb_e : G \to X$ extends continuously and G-equivariantly to an embedding of $\partial_{Gr} G$ that is a homeomorphism onto its image $\Lambda_e(G) \subset \partial_M X_e$.

We note that Theorem 4.4.7 and [DT, Proposition 3.2] imply that boundary convex cocompactness is invariant under quasi-isometric embeddings.

Remark 4.4.9 (The compactness assumption on $\Lambda_e(G)$ is essential for Theorem 4.4.8) Consider the group $G = \mathbb{Z}^2 * \mathbb{Z} * \mathbb{Z} = \langle a, b \rangle * \langle c \rangle * \langle d \rangle$ acting on its Cayley graph. Consider the subgroup $H = \langle a, b, c \rangle$. Since the H is isometrically embedded and convex in G, it follows that $\partial_M H_e \cong \Lambda_e(H) \subset \partial_M G_e$ and $\mathfrak{H}_e(H) = H$ for any $e \in G$, whereas H is not hyperbolic and thus not stable in G. In fact the compactness assumption ensures that the weak hull will be a subspace of some $X_e^{(N)}$, i.e., hyperbolic.

4.4.2 Height, width, bounded packing

Antolín, Mj, Sisto and Taylor in [AMST] use the boundary cocompactness characterization to extend some well-known intersection properties of quasi-convex subgroups of hyperbolic or relatively hyperbolic groups [GMRS, HW] to the context of stable subgroups of finitely generated groups:

Theorem 4.4.10 *Let $H_1, \dots H_l$ be stable subgroups of a finitely generated group. Then the collection $H = \{H_1, \dots, H_l\}$ satisfies the following:*

1 H has finite height.

2 H has finite width.
3 H has bounded packing.

In particular they show that if (G, H) is any group-subgroup pair satisfying one of the conditions in Theorem 4.4.1, then H has finite height, finite width, and bounded packing.

4.5 A metrisable topology on the Morse boundary

Cashen and Mackay have introduced a topology on the Morse boundary of a proper geodesic space that is first countable and regular and is metrisable when X is a group [CM]. They use a generalized notion of contracting geodesics which follows that of Arzhantseva, Cashen, Gruber, and Hume [ACGH1]. We present the definition here:

Definition 4.5.1 We call a function ρ *sublinear* if it is non-decreasing, eventually non-negative, and $\lim_{r\to\infty} \rho(r)/r = 0$.

Definition 4.5.2 Let $\gamma\colon [0,\infty) \to X$ be a geodesic ray in a proper geodesic metric space X, and let π_γ be the closest point projection to γ. Then, for a sublinear function ρ, we say that γ is *ρ-contracting* if for all x and y in X:

$$d(x,y) \le d(x,\gamma) \implies \operatorname{diam}(\pi_\gamma(x) \cup \pi_\gamma(y)) \le \rho(d(x,\gamma)).$$

We see that Definition 4.2.1 is simply this definition if we ask that ρ is the constant function D. We revisit the example in Remark 4.2.3: the space X was a ray Y with set of intervals $\{I_i\}$ of length i with an interval of length i^2 attached to the endpoints of I_i. We noted that this ray was not strongly contracting. It is not hard to see that it is $\sqrt{r}$-contracting with this more general definition. We showed in Remark 4.2.3 that this ray was Morse. This fact is no coincidence; in proper geodesic spaces, ρ-contracting is equivalent to being Morse [ACGH1].

Cashen and Mackay introduce a new topology on the contracting boundary that is finer than the "subspace topology" defined with the Gromov product but less fine than the direct limit topology. They call this topology the *topology of fellow-travelling quasi-geodesics* and denote it $\mathcal{FQ}$. The idea is that a geodesic α is close to a geodesic β if all quasi-geodesics tending to α closely fellow-travel quasi-geodesics tending to β for a long time.

One major difference from the direct limit topology on the Morse boundary is that rays with increasingly bad contracting functions can converge to a contracting ray. Recall again the space $\tilde{Y}$ from Example 4.2.6 and the collection of rays $\{\beta_i^j\}$ from Remark 4.2.7 . Consider the sequence $\{\beta_i^i\}$. It is not hard to see that in the topology of the fellow-travelling quasi-geodesics this converges to the geodesic ray $\alpha = aaaa\ldots$. The set $\{\beta_i^i\} \cup \{\alpha\}$ will be compact in $\mathcal{FQ}$. We note that the ray β_i^i has a constant contracting function $\rho_i = i$, so this compact set has arbitrarily bad contracting geodesics.

The $\mathcal{FQ}$ topology keeps many of the desirable properties of the Morse boundary with the direct limit topology. First, Cashen and Mackay show that it is a quasi-isometry invariant, first-countable and regular. Second, if in addition X is a group, then this boundary is second countable, and thus metrisable. Third, they prove a weak version of North–South dynamics for the action of a group on its contracting boundary as in Theorem 4.2.14 by Murray. Fourth, they show that if you restrict the $\mathcal{FQ}$ topology to rays that live in a single stratum and take the direct limit then this is homeomorphic to the Morse boundary. Finally, they show that if X is hyperbolic, then this boundary is homeomorphic to the Gromov boundary. In summary:

Theorem 4.5.3 ([CM]) *Let X be a proper geodesic space. The contracting boundary with the $\mathcal{FQ}$ topology, $\partial_c^{\mathcal{FQ}} X$, is*

1 a quasi-isometry invariant;

2 first countable and regular and if in in addition X is a group then $\partial_c^{\mathcal{FQ}} X$ is second countable and thus metrisable;

3 $\varinjlim_{\rho} \partial_c^{\mathcal{FQ}} X|_{\rho-contracting}$ is homeomorphic to the Morse boundary;

4 homeomorphic to the Gromov boundary if X is hyperbolic;

5 has weak North–South dynamics à la Theorem 4.2.14.

There are still many open questions about this topology: it is known that this topology is not in general homeomorphic to the subspace topology, but the example given is not a space with a geometric group action. Is this topology different from the subspace topology in the presence of a geometric group action? We know the space is metrisable, but can we give a useful description of a metric? If so, we can ask whether $\partial q \colon \partial_c^{\mathcal{FC}} X \to \partial_c^{\mathcal{FC}} Y$ is a quasi-symmetry whenever $q \colon X \to Y$ is a quasi-isometry?

Acknowledgements

The author would like to thank Ruth Charney and Mark Hagen for carefully reading drafts of this manuscript and offering many helpful suggestions. The author was partially supported at the Technion by a Zuckerman STEM Leadership Postdoctoral Fellowship and the Israel Science Foundation (Grant 1026/15).

References

[ACGH1] Goulnara N. Arzhantseva, Christopher H. Cashen, Dominik Gruber, and David Hume. Characterizations of morse quasi-geodesics via superlinear divergence and sublinear contraction. *Doc. Math.*, 22 (2017), 1193–1225.

[ACGH2] Goulnara N. Arzhantseva, Christopher H. Cashen, Dominik Gruber, and David Hume. Negative curvature in graphical small cancellation groups. To appear in *Groups Geom. Dyn.*

[AD] Goulnara N. Arzhantseva and Thomas Delzant. Examples of random groups. 2008.

[A-K] Yael Algom-Kfir. Strongly contracting geodesics in outer space. *Geom. Topol.*, 15(4):2181–2233, 2011.

[AMST] Yago Antolín, Mahan Mj, Alessandro Sisto, and Samuel J. Taylor. Intersection properties of stable subgroups and bounded cohomology. *Indiana Univ. Math. J.*, 68 (2019), 179–199.

[ADT] Tarik Aougab, Matthew Gentry Durham, and Samuel J Taylor. Pulling back stability with applications to $Out(F_n)$ and relatively hyperbolic groups (English summary). *J. Lond. Math. Soc.*, (2) 96 (2017), no. 3, 565–583.

[Ba] Werner Ballmann. *Lectures on spaces of nonpositive curvature*, volume 25 of *DMV Seminar*. Birkhäuser Verlag, Basel, 1995. With an appendix by Misha Brin.

[BaB] Werner Ballmann and Michael Brin. Orbihedra of nonpositive curvature. *Inst. Hautes Études Sci. Publ. Math.*, (82):169–209 (1996), 1995.

[Be] Jason A. Behrstock. Asymptotic geometry of the mapping class group and Teichmüller space. *Geom. Topol.*, 10:1523–1578, 2006.

[BC] Jason Behrstock and Ruth Charney. Divergence and quasimorphisms of right-angled Artin groups. *Math. Ann.*, 352(2):339–356, 2012.

[BDM] Jason Behrstock, Cornelia Druţu, and Lee Mosher. Thick metric spaces, relative hyperbolicity, and quasi-isometric rigidity. *Math. Ann.*, 344(3):543–595, 2009.

[BD] Jason Behrstock and Cornelia Druţu. Divergence, thick groups, and short conjugators. *Illinois J. Math.*, 58(4):939–980, 2014.

[BHS] Jason Behrstock, Mark F. Hagen, and Alessandro Sisto. Asymptotic dimension and small-cancellation for hierarchically hyperbolic spaces and groups. *Proc. Lond. Math. Soc.*, (3) 114 (2017), no. 5, 890–926.

[BeB] Mladen Bestvina and Ken Bromberg. On the asymptotic dimension of the curve complex. To appear in *Geom. Topol.*

[BBF] Mladen Bestvina, Ken Bromberg, and Koji Fujiwara. Constructing group actions on quasi-trees and applications to mapping class groups. *Publ. Math. Inst. Hautes Études Sci.*, 122:1–64, 2015.

[BeF] Mladen Bestvina and Koji Fujiwara. A characterization of higher rank symmetric spaces via bounded cohomology. *Geom. Funct. Anal.*, 19(1):11–40, 2009.

[BM] Mladen Bestvina and Geoffrey Mess. The boundary of negatively curved groups. *J. Amer. Math. Soc.*, 4(3):469–481, 1991.

[Bo1] Brian H. Bowditch. Cut points and canonical splittings of hyperbolic groups. *Acta Math.*, 180(2):145–186, 1998.

[Bo2] B.H. Bowditch. Peripheral splittings of groups. *Trans. Amer. Math. Soc.*, 353(10):4057–4082, 2001.

[Bo3] B.H. Bowditch. Relatively hyperbolic groups. *Internat. J. Algebra Comput.*, 22(3):1250016, 66, 2012.

[BH] Martin R. Bridson and André Haefliger. *Metric spaces of non-positive curvature*, volume 319 of *Grundlehren der Mathematischen Wissenschaften [Fundamental Principles of Mathematical Sciences]*. Springer-Verlag, Berlin, 1999.

[BrF] Jeffrey Brock and Benson Farb. Curvature and rank of Teichmüller space. *Amer. J. Math.*, 128(1):1–22, 2006.

[BMM] Jeffrey Brock, Howard Masur, and Yair Minsky. Asymptotics of Weil-Petersson geodesics II: bounded geometry and unbounded entropy. *Geom. Funct. Anal.*, 21(4):820–850, 2011.

[BS] Sergei Buyalo and Viktor Schroeder. *Elements of asymptotic geometry*. EMS Monographs in Mathematics. European Mathematical Society (EMS), Zürich, 2007.

[Bu] S.V. Buyalo. Asymptotic dimension of a hyperbolic space and the capacity dimension of its boundary at infinity. *Algebra i Analiz*, 17(2):70–95, 2005.

[BL] S.V. Buyalo and N. D. Lebedeva. Dimensions of locally and asymptotically self-similar spaces. *Algebra i Analiz*, 19(1):60–92, 2007.

[CF] Pierre-Emmanuel Caprace and Koji Fujiwara. Rank-one isometries of buildings and quasi-morphisms of Kac-Moody groups. *Geom. Funct. Anal.*, 19(5):1296–1319, 2010.

[Ca] Christopher H. Cashen. Quasi-isometries need not induce homeomorphisms of contracting boundaries with the Gromov product topology. *Anal. Geom. Metr. Spaces*, 4:278–281, 2016.

[CM] C. H. Cashen and J. M. Mackay. A metrizable topology on the contracting boundary of a group. To appear in *Trans. Amer. Math. Soc.*

[ChD] Ruth Charney and Michael W. Davis. Finite $K(\pi,1)$s for Artin groups. In *Prospects in topology (Princeton, NJ, 1994)*, volume

138 of *Ann. of Math. Stud.*, pages 110–124. Princeton Univ. Press, Princeton, NJ, 1995.

[CMC] Ruth Charney, Devin Murray, and Matthew Cordes. Quasi-mobius homeomorphisms of the morse boundary. To appear in *Bull. London Math. Soc.*

[CS1] Ruth Charney and Harold Sultan. Contracting boundaries of CAT(0) spaces. *J. Topol*, 8 (2015), no. 1, 93–117.

[CS2] Ruth Charney and Harold Sultan. Contracting boundaries of CAT(0) spaces. *J. Topol.*, 8(1):93–117, 2015.

[CoD] Matthew Cordes and Matthew Gentry Durham. Boundary convex cocompactness and stability of subgroups of finitely generated groups. to appear in *Int. Math. Res. Notices.*

[CH1] Matthew Cordes and David Hume. Relatively hyperbolic groups with fixed peripherals. To appear in *Isr. J. Math.*

[CH2] Matthew Cordes and David Hume. Stability and the Morse boundary. *J. Lond. Math. Soc. (2)*, 95(3):963–988, 2017.

[Co] Matthew Cordes. Morse boundaries of proper geodesic metric spaces (English summary). *Groups, Geometry, and Dynamics*, 11 (2017), no. 4, 1281–1306.

[CK] Christopher B. Croke and Bruce Kleiner. Spaces with nonpositive curvature and their ideal boundaries. *Topology*, 39(3):549–556, 2000.

[DGO] F. Dahmani, V. Guirardel, and D. Osin. Hyperbolically embedded subgroups and rotating families in groups acting on hyperbolic spaces. *Mem. Amer. Math. Soc.*, 245 (2017), no. 1156, v + 152 pp.

[DKL] Spencer Dowdall, Autumn Kent, and Christopher J. Leininger. Pseudo-Anosov subgroups of fibered 3-manifold groups. *Groups Geom. Dyn.*, 8(4):1247–1282, 2014.

[DMS] Cornelia Druţu, Shahar Mozes, and Mark Sapir. Divergence in lattices in semisimple Lie groups and graphs of groups. *Trans. Amer. Math. Soc.*, 362(5):2451–2505, 2010.

[DS] Cornelia Druţu and Mark Sapir. Tree-graded spaces and asymptotic cones of groups. *Topology*, 44(5):959–1058, 2005. With an appendix by Denis Osin and Sapir.

[DT] Matthew Gentry Durham and Samuel J. Taylor. Convex cocompactness and stability in mapping class groups. *Algebr. Geom. Topol.*, 15(5):2839–2859, 2015.

[F] B. Farb. Relatively hyperbolic groups. *Geom. Funct. Anal.*, 8(5):810–840, 1998.

[FM] Benson Farb and Lee Mosher. Convex cocompact subgroups of mapping class groups. *Geom. Topol.*, 6:91–152 (electronic), 2002.

[GdlH] Étienne Ghys and Pierre de la Harpe. Quasi-isométries et quasi-géodésiques. In *Sur les groupes hyperboliques d'après Mikhael Gromov (Bern, 1988)*, volume 83 of *Progr. Math.*, pages 79–102. Birkhäuser Boston, Boston, MA, 1990.

[GMRS] Rita Gitik, Mahan Mitra, Eliyahu Rips, and Michah Sageev. Widths of subgroups. *Trans. Amer. Math. Soc.*, 350(1):321–329, 1998.

[Grof] Bradley W. Groff. Quasi-isometries, boundaries and JSJ-decompositions of relatively hyperbolic groups. *J. Topol. Anal.*, 5(4):451–475, 2013.

[Grom1] M. Gromov. Asymptotic invariants of infinite groups. In *Geometric group theory, Vol. 2 (Sussex, 1991)*, volume 182 of *London Math. Soc. Lecture Note Ser.*, pages 1–295. Cambridge Univ. Press, Cambridge, 1993.

[Grom2] M. Gromov. Random walk in random groups. *Geom. Funct. Anal.*, 13(1):73–146, 2003.

[Hag] Mark F. Hagen. Weak hyperbolicity of cube complexes and quasi-arboreal groups. *J. Topol.*, 7(2):385–418, 2014.

[Ham1] Ursula Hamenstädt. Word hyperbolic extensions of surface groups. *ArXiv e-prints*, 2005.

[Ham2] Ursula Hamenstädt. Rank-one isometries of proper CAT(0)-spaces. In *Discrete groups and geometric structures*, volume 501 of *Contemp. Math.*, pages 43–59. Amer. Math. Soc., Providence, RI, 2009.

[HH] Ursula Hamenstädt and Sebastian Hensel. Convex cocompact subgroups of Out($\mathbb{F}$). *ArXiv e-prints*, 2014.

[HLS] N. Higson, V. Lafforgue, and G. Skandalis. Counterexamples to the Baum-Connes conjecture. *Geom. Funct. Anal.*, 12(2):330–354, 2002.

[HK] G. Christopher Hruska and Bruce Kleiner. Hadamard spaces with isolated flats. *Geom. Topol.*, 9:1501–1538, 2005. With an appendix by the authors and Mohamad Hindawi.

[HW] G. Christopher Hruska and Daniel T. Wise. Packing subgroups in relatively hyperbolic groups. *Geom. Topol.*, 13(4):1945–1988, 2009.

[KL1] Autumn Kent and Christopher J. Leininger. Subgroups of mapping class groups from the geometrical viewpoint. In *In the tradition of Ahlfors-Bers. IV*, volume 432 of *Contemp. Math.*, pages 119–141. Amer. Math. Soc., Providence, RI, 2007.

[KL2] Autumn Kent and Christopher J. Leininger. Shadows of mapping class groups: capturing convex cocompactness. *Geom. Funct. Anal.*, 18(4):1270–1325, 2008.

[KLS] Autumn Kent, Christopher J. Leininger, and Saul Schleimer. Trees and mapping class groups. *J. Reine Angew. Math.*, 637:1–21, 2009.

[KMT] Thomas Koberda, Johanna Mangahas, and Samuel J. Taylor. The geometry of purely loxodromic subgroups of right-angled artin groups. *ArXiv e-prints*, 2014.

[LS] Christopher Leininger and Saul Schleimer. Hyperbolic spaces in Teichmüller spaces. *J. Eur. Math. Soc. (JEMS)*, 16(12):2669–2692, 2014.

[MS] John M. Mackay and Alessandro Sisto. Embedding relatively hyperbolic groups in products of trees. *Algebr. Geom. Topol.*, 13(4):2261–2282, 2013.

[MM1] Howard A. Masur and Yair N. Minsky. Geometry of the complex of curves. I. Hyperbolicity. *Invent. Math.*, 138(1):103–149, 1999.

[MM2] H. A. Masur and Y. N. Minsky. Geometry of the complex of curves. II. Hierarchical structure. *Geom. Funct. Anal.*, 10(4):902–974, 2000.

[Min] Honglin Min. Hyperbolic graphs of surface groups. *Algebr. Geom. Topol.*, 11(1):449–476, 2011.

[Mins] Yair N. Minsky. Quasi-projections in Teichmüller space. *J. Reine Angew. Math.*, 473:121–136, 1996.

[Mo] Harold Marston Morse. A fundamental class of geodesics on any closed surface of genus greater than one. *Trans. Amer. Math. Soc.*, 26(1):25–60, 1924.

[Mu] Devin Murray. Topology and dynamics of the contracting boundary of cocompact CAT(0) spaces. *ArXiv e-prints*, 2015.

[OOS] Alexander Yu. Ol'shanskii, Denis V. Osin, and Mark V. Sapir. Lacunary hyperbolic groups. *Geom. Topol.*, 13(4):2051–2140, 2009. With an appendix by Michael Kapovich and Bruce Kleiner.

[Osa] D. Osajda. Small cancellation labellings of some infinite graphs and applications. *ArXiv e-prints*, June 2014.

[Osi1] D. Osin. Asymptotic dimension of relatively hyperbolic groups. *Int. Math. Res. Not.*, (35):2143–2161, 2005.

[Osi2] Denis V. Osin. Relatively hyperbolic groups: intrinsic geometry, algebraic properties, and algorithmic problems. *Mem. Amer. Math. Soc.*, 179(843):vi+100, 2006.

[Osi3] D. Osin. Acylindrically hyperbolic groups. *Trans. Amer. Math. Soc.*, 368(2):851–888, 2016.

[Pan] Pierre Pansu. Dimension conforme et sphère à l'infini des variétés à courbure négative. *Ann. Acad. Sci. Fenn. Ser. A I Math.*, 14(2):177–212, 1989.

[Pau] Frédéric Paulin. Un groupe hyperbolique est déterminé par son bord. *J. London Math. Soc. (2)*, 54(1):50–74, 1996.

[Q] Yulan Qing. *Boundary of CAT(0) Groups With Right Angles.* ProQuest LLC, Ann Arbor, MI, 2013. Thesis (Ph.D.)–Tufts University.

[R] John Roe. Hyperbolic groups have finite asymptotic dimension. *Proc. Amer. Math. Soc.*, 133(9):2489–2490 (electronic), 2005.

[Si] Alessandro Sisto. Quasi-convexity of hyperbolically embedded subgroups. *Mathematische Zeitschrift*, pages 1–10, 2016.

[Su] Harold Sultan. Hyperbolic quasi-geodesics in CAT(0) spaces. *Geom. Dedicata*, 169:209–224, 2014.

[Sw] G.A. Swarup. On the cut point conjecture. *Electron. Res. Announc. Amer. Math. Soc.*, 2(2):98–100 (electronic), 1996.

[T] Hung Cong Tran. Divergence spectra and Morse boundaries of relatively hyperbolic groups. *ArXiv e-prints*, 2017.

[Wi] Julia M. Wilson. A CAT(0) group with uncountably many distinct boundaries. *J. Group Theory*, 8(2):229–238, 2005.

[Wr] Nick Wright. Finite asymptotic dimension for CAT(0) cube complexes. *Geom. Topol.*, 16(1):527–554, 2012.

[Y] Guoliang Yu. The Novikov conjecture for groups with finite asymptotic dimension. *Ann. of Math. (2)*, 147(2):325–355, 1998.

5
What is a hierarchically hyperbolic space?

Alessandro Sisto
Department of Mathematics
ETH Zürich
Rmistrasse 101
8092 Zürich
Switzerland

Abstract

The first part of this survey is a heuristic, non-technical discussion of what an HHS is, and the aim is to provide a good mental picture both to those actively doing research on HHSs and to those who only seek a basic understanding out of pure curiosity. It can be read independently of the second part, which is a detailed technical discussion of the axioms and the main tools for dealing with HHSs.

Introduction

Hierarchically hyperbolic spaces (HHSs) were introduced in [BHS2] as a common framework to study mapping class groups and cubical groups. The definition is inspired by the extremely successful Masur-Minsky machinery for studying mapping class groups [MM, MM2, Be]. Since [BHS2], the list of examples has expanded significantly [BHS1, BHS3, HS], and the HHS framework has been used to prove several new results, including new results for mapping class groups and cubical groups. For example, the previously known bound for the asymptotic dimension of mapping class groups has been dramatically improved in [BHS3], while the main result from [BHS4] is that top-dimensional quasi-flats in HHSs stay within bounded distance from a finite union of "standard orthants", a fact that was known neither for mapping class groups nor for cubical groups without imposing additional constraints (see [Hu]).

The aim of this survey article, however, is not to present the state of the art of the field, which is very much evolving. In this direction, we only give a brief description of all the relevant papers and the known examples of HHSs below. The main aim of this survey is, instead, to discuss the geometry of HHSs, only assuming that the reader is familiar with (Gromov-)hyperbolic spaces. The definition of HHS is admittedly hard to digest if one is not presented with the geometric intuition behind it, and the aim of this survey is to remedy this shortcoming. The first part is aimed at the casual reader and gives a general idea of what an HHS looks like. We will discuss the various notions in the main motivating examples too; the reader can use whichever example they are familiar with to gain better understanding.

The second part of the survey is mainly aimed at those who want to do research on HHSs, as well as to those who seek deeper understanding. We will discuss every axiom in detail, and then we will proceed to discuss the main tools one can use to study HHSs. Anyone who becomes familiar with the material that will be presented will have a rather deep understanding of HHSs. And will be ready to tackle one of the many open questions asked in the papers we describe below...

State of the art and known examples

Here is a list of groups and space known to be hierarchically hyperbolic at the time of writing.

- Hyperbolic spaces.
- Direct products of HHSs [BHS1, Proposition 8.25].
- Spaces hyperbolic relative to (uniform) HHSs [BHS1, Theorem 9.1].
- Mapping class groups of finite-type surfaces. See [BHS2, Theorem G] and references therein.
- Teichmüller spaces of finite-type surfaces endowed with either the Teichmüller or the Weil–Petersson metric. See [BHS2, Theorem G] and references therein.
- Many groups acting geometrically on CAT(0) cube complexes [HS, Theorem A].
- Fundamental groups of graphs of groups satisfying suitable conditions [BHS1, Theorem 8.6, Corollary 8.22], including fundamental groups of non-geometric graph manifolds and mixed 3-manifolds [BHS1, Theorem 10.1] and others [S1, S2].
- Dehn-filling/small-cancellation-type quotients of hierarchically hyperbolic groups [BHS3, Theorem E].

Here is a brief description of the relevant papers.

- In [BHS2], J. Behrstock, M. Hagen and I axiomatised the Masur-Minsky machinery, extended it to right-angled Artin groups (and many other groups acting on CAT(0) cube complexes including fundamental groups of special cube complexes), and initiated the study of the geometry of hierarchically hyperbolic groups by studying quasi-flats via "coarse differentiation".
- In [BHS1] we simplified the list of axioms, which allowed us to extend the list of examples of hierarchically hyperbolic groups (and to significantly simplify the Masur-Minsky approach). The paper contains a combination theorem for trees of HHSs, as well as other results to construct new HHSs out of old ones.
- Speaking of new examples, in [HS] M. Hagen and T. Susse prove that there are many more CAT(0) cube complexes that are HHSs.
- [BHS3] deals with asymptotic dimension. We show finiteness of the asymptotic dimension of hierarchically hyperbolic groups, giving explicit estimates in certain cases. In the process we drastically improve previously known bounds on the asymptotic dimension of mapping class groups. We also show that many natural (small cancellation) quotients of hierarchically hyperbolic groups are hierarchically hyperbolic.
- In [DHS], M. Durham, M. Hagen and I introduce a compactification of hierarchically hyperbolic groups, related to Thurston's compactification of Teichmüller space in the case of mapping class groups and Teichmüller space. This compactification turns out to be very well-behaved as, for example, "quasi-convex subgroups" in a suitable sense have a well-defined and easily recognisable limit set (while the situation for Teichmüller space is more complicated). Constructing this compactification allowed us, for example, to study dynamical properties of individual elements and to prove a rank rigidity result.
- Further study of the HHS boundary is carried out in [M1, M2], where S. Mousley shows non-existence of boundary maps in certain cases and other exotic phenomena.
- In [BHS4], J. Behrstock, M. Hagen and I study the geometry of quasi-flats, that is to say images of quasi-isometric embeddings of $\mathbb{R}^n$ in hierarchically hyperbolic spaces. More specifically, we show that top-dimensional quasi-flats lie within finite distance of a union of "standard orthants". This simultaneously solves open questions and conjectures for most of the motivating examples of hierarchically hyperbolic groups, for example a conjecture of B. Farb for mapping class groups, and one by J. Brock for the Weil–Petersson metric, and it is new in the context of CAT(0) cube complexes too.

- In [S1, S2], D. Spriano studies non-trivial HHS structures on hyperbolic spaces, and uses them to show that certain natural amalgamated products of hierarchically hyperbolic groups are hierarchically hyperbolic.
- In [ABD], C. Abbott, J. Behrstock and M. Durham prove that hierarchically hyperbolic groups admit a "best" acylindrical action on a hyperbolic space, and provide a complete classification of stable subgroups of hierarchically hyperbolic groups.
- In [Hae], T. Haettel studies homomorphisms from higher rank lattices to hierarchically hyperbolic groups, finding severe restrictions.
- In [ST], using part of an HHS structure, S. Taylor and I studied various notions of projections, including subsurface projection for mapping class groups, along a random walk, and used this to prove a conjecture of I. Rivin on random mapping tori.

Acknowledgements

This article has been written for the proceedings for the "Beyond hyperbolicity" conference held in June 2016 in Cambridge, UK. The author would like to thank Mark Hagen, Richard Webb, and Henry Wilton for organising the conference and the wonderful time he had in Cambridge.

The author would also like to thank Jason Behrstock, Mark Hagen, Davide Spriano, and an anonymous referee for useful comments on previous drafts of this survey.

5.1 Heuristic discussion

5.1.1 Standard product regions

In this section we discuss the first heuristic picture of an HHS, which is the one provided by *standard product regions.*

If an HHS $\mathcal{X}$ is not hyperbolic, then the obstruction to its hyperbolicity is encoded by the collection of its standard product regions. These are quasi-isometrically embedded subspaces that split as direct products, and the crucial fact is that each standard product region, as well as each of its factors, is an HHS itself, and in fact an HHS of lower "complexity". It is not very important at this point, but the complexity is roughly speaking the length of a longest chain of standard product regions $P_1 \subsetneq P_2 \subsetneq \cdots \subsetneq P_n$ contained in the HHS; what is important right now is that factors of standard product regions are "simpler" HHSs, and the "simplest" HHSs are hyperbolic spaces. This is what allows for induction arguments, where the base case is that of hyperbolic spaces.

Standard product regions encode entirely the non-hyperbolicity of the HHS $\mathcal{X}$ in the following sense. Given a, say, length metric space (Z, d) and a collection of subspaces $\mathcal{P}$, one can define the cone-off of Z with respect to the collection of subspaces (in several different ways that coincide up to quasi-isometry, for example) by setting $d'(x,y) := 1$ for all x, y contained in the same $P \in \mathcal{P}$ and $d'(x,y) = d(x,y)$ otherwise, and declaring the cone-off distance between two points x, y to be $\inf_{x=x_0,\dots,x_n=y} \sum d'(x_i, x_{i+1})$. This has the effect of collapsing all $P \in \mathcal{P}$ to bounded sets, and the reason that this is a sensible thing to do is that one might want to consider the geometry of Z "up to" the geometry of the $P \in \mathcal{P}$. When Z is a graph, as is most often the case for us, coning-off amounts to adding edges connecting pairs of vertices contained in the same $P \in \mathcal{P}$.

Back to HHSs: when coning-off all standard product regions of an HHS one obtains a hyperbolic space, which we denote $\mathcal{C}S$.[1] In other words, an HHS is weakly hyperbolic relative to the standard product regions. Roughly speaking, when moving around $\mathcal{X}$ one is either moving in the hyperbolic space $\mathcal{C}S$ or in one of the standard product regions. The philosophy behind many induction arguments for HHSs is that when studying a certain "phenomenon", either it leaves a visible trace in $\mathcal{C}S$, or it is "confined" in a standard product region, and can hence be studied there. For example, if the HHS is in fact a group, one can consider the subgroup generated by an element g, and it turns out that either the orbit maps of g in $\mathcal{C}S$ are quasi-isometric embeddings, or g virtually fixes a standard product region [DHS].

So far we discussed the "top-down" point of view on standard product regions, but there is also a "bottom-up" approach. In fact, one can regard HHSs as built up inductively starting from hyperbolic spaces, in the following way:

- hyperbolic spaces are HHSs,
- direct products of HHSs are HHSs,
- "hyperbolic-like" arrangements of HHSs are HHSs.

The third bullet refers to $\mathcal{C}S$ being hyperbolic, and the fact that $\mathcal{C}S$ can also be thought of as encoding the intersection pattern of standard product regions. Incidentally, I believe that there should be a characterisation of HHSs that looks like the list above, i.e. that by suitably formalising the third bullet one can obtain a characterisation of HHSs.

[1] This notation is taken from the mapping class group context, even though it's admittedly not the best notation in other examples.

This has not been done yet, though. There is, however, a combination theorem for trees of HHSs in this spirit [BHS1].

One final thing to mention is that standard product regions have well-behaved coarse intersections, meaning that the coarse intersection of two standard product regions is well-defined and coarsely coincides with some standard product region. In other words, $\mathcal{X}$ is obtained by gluing together standard product regions along sub-HHSs, so a better version of the third bullet above would be "hyperbolic-like arrangements of HHS glued along sub-HHSs are HHSs".

In the examples

We now discuss standard product regions in motivating examples of HHSs.

RAAGs. Consider a simplicial graph Γ. Whenever one has a (full) subgraph Λ of Γ which is the join of two (full, non-empty) subgraphs Γ_1, Γ_2, then the RAAG A_Γ contains an undistorted copy of the RAAG $A_\Lambda \approx A_{\Lambda_1} \times A_{\Lambda_2}$. Such subgroups and their cosets are the standard product regions of A_Γ. In this case, $\mathcal{C}S$ is a Cayley graph of A_Γ with respect to an infinite generating set (unless Γ consists of a single vertex), namely $V\Gamma \cup \{A_\Lambda < A_\Gamma : \Lambda = join(\Lambda_1, \Lambda_2)\}$. A given HHS can be given different HHS structures (which turns out to allow for more flexibility when performing various constructions, rather than being a drawback), and one instance of this is that one can regard as standard product regions all A_Λ where Λ is any proper subgraph of Γ, one of the factors being trivial. In this case $\mathcal{C}S$ is the Cayley graph of A_Γ with respect to the generating set $V\Gamma \cup \{A_\Lambda < A_\Gamma : \Lambda \text{ proper subgraph of } \Gamma\}$, which is perhaps more natural.

For both HHS structures described above, $\mathcal{C}S$ is not only hyperbolic, but in fact quasi-isometric to a tree.

Notice that other quasi-trees associated to RAAGs had been previously considered in the literature, namely contact graphs [Hag], which can actually be associated to a general CAT(0) cube complex, and extension graphs [KK], as well as Bass–Serre trees of natural splittings. For example, all the aforementioned graphs can be shown to prove acylindrical hyperbolicity of RAAGs that do not split as products (see [MO] for the Bass–Serre tree case).

Mapping class groups. Given a surface S, there are some "obvious" subgroups of $MCG(S)$ that are direct products. In fact, consider two disjoint (essential) subsurfaces Y, Z of S. Any two self-homeomorphisms of S supported respectively on Y and Z commute. This yields (up to ignoring issues related to the difference between boundary components and punctures that I do not want to get into) a subgroup of $MCG(S)$ isomorphic to $MCG(Y) \times MCG(Z)$. Such subgroups are in fact undistorted. One can similarly consider finitely many disjoint subsurfaces instead, and this yields the standard product regions in $MCG(S)$. More precisely, one should fix representatives of the (finitely many) topological types of collections of disjoint subsurfaces, and consider the cosets of the subgroups as above. In terms of the marking graph, product regions are given by all markings containing a given sub-marking.

In this case it shouldn't be too hard to convince oneself that $\mathcal{C}S$ as defined above is quasi-isometric to the curve complex, see [MM]. To re-iterate the philosophy explained above, if some behaviour within $MCG(S)$ is not confined to a proper subsurface Y, then the geometry of $\mathcal{C}S$ probably comes into play when studying it, and otherwise it is most convenient to study the problem on the simpler subsurface Y.

CAT(0) cube complexes. Hyperplanes are crucial for studying CAT(0) cube complexes, and the carrier of a hyperplane (meaning the union of all cubes that the hyperplane goes through) is naturally a product of the hyperplane and an interval. It is then natural, when trying to define an HHS structure on a CAT(0) cube complex, to regard carriers of hyperplanes as standard product regions, even though one of the factors is bounded.

This is not enough, though. As mentioned above, coarse intersections of standard product regions should be standard product regions. But it is easy to describe the coarse intersection of two carriers of hyperplanes, or more generally the coarse intersection of two convex subcomplexes. Given a convex subcomplex Y of the CAT(0) cube complex $\mathcal{X}$, one can consider the *gate map* $\mathfrak{g}_Y : \mathcal{X} \to Y$, which is the closest-point projection in either the CAT(0) or the ℓ^1–metric (they coincide). More combinatorially, for $x \in \mathcal{X}^{(0)}$, $\mathfrak{g}_Y(x)$ is defined by the property that the hyperplanes separating x from $\mathfrak{g}_Y(x)$ are exactly those separating x from Y. The coarse intersection of convex subcomplexes Y, Z is just $\mathfrak{g}_Y(Z)$, which is itself a convex subcomplex.

Back to constructing an HHS structure on a CAT(0) cube complex: we now know that we need to include as standard product regions all gates of carriers in other carriers. But then we are still not done, because for the same reason that we need to take gates of carriers we also need to take gates of gates, and so on. Also, we need this process to stabilise eventually (which is not always the case, unfortunately), because an HHS needs to have finite complexity to allow for induction arguments. All these considerations lead to the definition of *factor system*. Rather than carriers of hyperplanes, we will consider combinatorial hyperplanes, which are the two copies of a hyperplane that bound its carrier, but this does not make a substantial difference for the purposes of this discussion.

Definition 5.1.1 A factor system $\mathcal{F}$ for the cube complex $\mathcal{X}$ is a collection of convex subcomplexes so that:

1 all combinatorial hyperplanes are in $\mathcal{F}$;
2 there exists $\xi \geq 0$ so that if $F, F' \in \mathcal{F}$ and $\mathfrak{g}_F(F')$ has diameter at least ξ, then $\mathfrak{g}_F(F') \in \mathcal{F}$;
3 $\mathcal{F}$ is uniformly locally finite.

Any factor system $\mathcal{F}$ on a cube complex gives an HHS structure, where $\mathcal{C}S$ is obtained by coning off all members of $\mathcal{F}$. It turns out that $\mathcal{C}S$ is quasi-isometric to a tree. It is proven in [HS] that many (and all known) CAT(0) cube complexes admitting a proper cocompact action by isometries have a factor system, and are therefore HHSs.

5.1.2 Projections to hyperbolic spaces

In this section we discuss a point of view on HHSs that is more similar to the actual definition, and it is in terms of "coordinates" in certain hyperbolic spaces.

We already saw that any HHS $\mathcal{X}$ comes equipped with a hyperbolic space, $\mathcal{C}S$, obtained by collapsing the standard product regions. In particular, there is a coarsely-Lipschitz map $\pi_S : \mathcal{X} \to \mathcal{C}S$.

The (coarse geometry of the) hyperbolic space $\mathcal{C}S$ is not enough to recover the whole geometry of $\mathcal{X}$, since it does not contain information about the standard product regions themselves. Hence, we want something to keep track of the geometry of the standard product regions. Since factors of standard product regions are HHSs themselves, they also come with a hyperbolic space obtained by collapsing the standard product sub-regions. Considering all standard product regions, we obtain by a collection of hyperbolic spaces $\{\mathcal{C}Y\}_{Y \in \mathfrak{S}}$, which, together, control the geometry of $\mathcal{X}$, as we are about to discuss. The index set $\mathfrak{S}$ is the set of factors of standard product regions, where the whole of $\mathcal{X}$ should be considered as a product region with (a trivial factor and the other factor being) $S \in \mathfrak{S}$, so as to include $\mathcal{C}S$ among the hyperbolic spaces we consider.[2]

Another piece of data we need is a collection of coarsely-Lipschitz maps $\pi_Y : \mathcal{X} \to \mathcal{C}Y$ for all $Y \in \mathfrak{S}$, which allow us to talk about the geometry of $\mathcal{X}$ "from the point of view of $\mathcal{C}Y$". These projection maps come from natural coarse retractions of $\mathcal{X}$ onto the standard product regions, composed with the collapsing maps, but for now it only matters that the π_Y exist.

Distance formula and hierarchy paths

The first way in which the $\mathcal{C}Y$ control the geometry of $\mathcal{X}$ is that whenever $x, y \in \mathcal{X}$ are far away, then their projections to some $\mathcal{C}Y$ are far away, so that any coarse-geometric feature of $\mathcal{X}$ leaves a trace in at least one of the $\mathcal{C}Y$. In fact, there is much better control on distances in $\mathcal{X}$ in terms of distances in the various $\mathcal{C}Y$, and this is given by the *distance formula*. This is perhaps the most important piece of machinery in the HHS world, and certainly the most iconic. To state it, we need a little bit of notation. We write $A \approx_K B$ if $A/K - K \leq B \leq KB + K$, and declare $\{\!\{A\}\!\}_L = A$ if $A \geq L$, and $\{\!\{A\}\!\}_L = 0$ otherwise. The distance formula says that for all sufficiently large L there exists K so that

$$\mathsf{d}_{\mathcal{X}}(x, y) \approx_K \sum_{Y \in \mathfrak{S}} \{\!\{\mathsf{d}_{\mathcal{C}Y}(\pi_Y(x), \pi_Y(y))\}\!\}_L$$

for all $x, y \in \mathcal{X}$. In words, the distance in $\mathcal{X}$ between two points is, up to multiplicative and additive constants, the sum of the distances between their far-away projections in the various $\mathcal{C}Y$. Very imprecisely, this is saying that $\mathcal{X}$ quasi-isometrically embeds in $\prod_{Y \in \mathfrak{S}} \mathcal{C}Y$ endowed with some sort of ℓ^1 metric. To save notation one usually writes $\mathsf{d}_Y(x, y)$ instead of $\mathsf{d}_{\mathcal{C}Y}(\pi_Y(x), \pi_Y(y))$.

Another important fact related to the distance formula is the existence of *hierarchy paths*, that is to say quasi-geodesics in $\mathcal{X}$ that shadow geodesics in each $\mathcal{C}Y$. Namely, there exists D so that for any $x, y \in \mathcal{X}$ there exists a (D, D)–quasigeodesic γ joining them so that $\pi_Y \circ \gamma$ is an unparametrised (D, D)–quasigeodesic in $\mathcal{C}Y$. Since $\mathcal{C}Y$ is hyperbolic, being an unparametrised quasigeodesic means that (the image of) $\pi_Y \circ \gamma$ is Hausdorff-close to a geodesic, and it "traverses" the geodesic coarsely monotonically. In most cases it is much better to deal with hierarchy paths than with geodesics.

[2] As a technical note, in the formal setup in Part 5.2, the index set $\mathfrak{S}$ is in bijection with the set of standard product regions, not necessarily identical to it. The factors of the standard product regions will then be denoted by F_Y, E_Y for $Y \in \mathfrak{S}$.

Consistency

The distance formula is not alone enough for almost anything, but the point is that it comes with a toolbox that one uses to control the various projection terms by constraining certain projections in terms of certain other projections. In Part 5.2, we will analyse these tools in detail. For now we will instead describe what happens to various projections when moving along a hierarchy path, which gives the right picture about how projections are constrained.

The short version of this subsection is: for certain pairs $Y, Z \in \mathfrak{S}$, along a hierarchy path one can only change the projections to $\mathcal{C}Y, \mathcal{C}Z$ in a specified order. This is sufficient to understand most of the next subsection.

Nesting. Let $x, y \in \mathcal{X}$ and suppose that $\mathsf{d}_Y(x,y)$ is large. Notice that Y (which is a factor of a standard product region) gives a bounded set in $\mathcal{C}S$ (which was obtained from $\mathcal{X}$ by collapsing standard product regions), which we denote ρ^Y_S. We know that when moving along any hierarchy path γ from x to y, the projection to $\mathcal{C}Y$ needs to change. This is how this happens: γ has an initial subpath where the projection to $\mathcal{C}Y$ coarsely does not change, while the projection to $\mathcal{C}S$ approaches ρ^Y_S. All the progress that needs to be made by γ in $\mathcal{C}Y$ is made by a middle subpath whose projection to $\mathcal{C}S$ remains close to ρ^Y_S. Then, there is a final subpath that does not make any progress in $\mathcal{C}Y$ and takes us from ρ^Y_S to $\pi_S(y)$ in $\mathcal{C}S$. In short, you can only make progress in $\mathcal{C}Y$ if you are close to ρ^Y_S in $\mathcal{C}S$.

The description above applies to more general pairs of elements of $\mathfrak{S}$, namely whenever S is replaced by some Z so that Y is properly *nested into* Z, denoted $Y \sqsubset Z$. Nesting just means that the factor Y is contained in the factor Z, or more precisely that there is a copy of Y in a standard product region that is contained in a copy of Z.

Orthogonality. We just saw that changing projections in both Y and Z when $Y \sqsubset Z$ can only be done in a rather specific way. The opposite situation is when Y and Z (which, recall, are factors of standard product regions) are *orthogonal*, denoted $Y \perp Z$, meaning that they are (contained in) different factors of the same product region. In this case, along a hierarchy path there is no constraint regarding which projection needs to change first, and in fact they can also change simultaneously. Orthogonality is what creates non-hyperbolic behaviour in HHSs, and is what one has to constantly fight against.

Transversality. When Y, Z are neither $\sqsubseteq$- nor $\perp$-comparable, we say that they are *transverse*. This is the generic case. When $Y \pitchfork Z$ and $x, y \in \mathcal{X}$ are so that $\mathsf{d}_Y(x,y), \mathsf{d}_Z(x,y)$ are both large, up to switching Y, Z what happens is the following. When moving along any hierarchy path from x to y one has to first change the projection to $\mathcal{C}Y$ until it coarsely coincides with $\pi_Y(z)$, and only then the projection to $\mathcal{C}Z$ can start moving from $\pi_Z(x)$ to $\pi_Z(y)$.

Arguably the most useful feature of transversality is a slight generalisation of this. Given $x, y \in \mathcal{X}$, and a set $\mathcal{Y} \subseteq \mathfrak{S}$ of pairwise transverse elements so that $\mathsf{d}_Y(x,y)$ is large for every $Y \in \mathcal{Y}$, there is a total order on $\mathcal{Y}$ so that, whenever $Y < Z$, along any hierarchy path from x to y the projection to $\mathcal{C}Y$ has to change before the projection to $\mathcal{C}Z$ does, as described above.

Realisation

Even though we did not formally describe them, we saw that for certain pairs $Y, Z \in \mathfrak{S}$, namely when $Y \not\perp Z$, there are some constraints on the projections of points in $\mathcal{X}$ to $\mathcal{C}Y, \mathcal{C}Z$. These are called *consistency inequalities*. As it turns out, the consistency inequalities are the only obstructions for "coordinates" $(b_Y \in \mathcal{C}Y)_{Y \in \mathfrak{S}}$ to be coarsely realised by a point x in $\mathcal{X}$, meaning that $\pi_Y(x)$ coarsely coincides with b_Y in each $\mathcal{C}Y$. This is important because it allows us to perform constructions in each of the $\mathcal{C}Y$ separately and then put everything back together.

To make this principle clear, we now give an example of a construction of this type. Say we want to construct a "coarse median" map $m : \mathcal{X}^3 \to \mathcal{X}$ (in the sense of [Bo1]), which let's just take to mean a coarsely-Lipschitz map so that $m(x,x,y)$ is coarsely x. Consider x, y, z in $\mathcal{X}$, and let us define $m(x,y,z)$ by defining its coordinates in the $\mathcal{C}Y$. Given $Y \in \mathfrak{S}$, the triangle with vertices $\pi_Y(x), \pi_Y(y), \pi_Y(z)$ has a coarse centre b_Y, because $\mathcal{C}Y$ is hyperbolic. It turns out that the coordinates (b_Y) satisfy the consistency inequalities, so that one can define $m(x,y,z)$ as the realisation point. As an aside, it is a nice exercise to use the properties of m to show that $\mathcal{X}$ satisfies a quadratic isoperimetric inequality.

To sum up, the distance formula says that the natural map $\mathcal{X} \to \prod_{Y \in \mathfrak{S}} \mathcal{C}Y$ is "coarsely injective", and the consistency inequalities provide a coarse characterisation of the image.

In the examples

RAAGs. In the case of RAAGs, $\mathfrak{S}$ (the set of factors of product regions) is the set of cosets of sub-RAAGs, considered up to *parallelism*. We say that $gA_\Lambda, hA_\Lambda \subseteq A_\Gamma$ are parallel if $g^{-1}h$ commutes with every element of A_Λ, which essentially means that there's a product $g(A_\Lambda \times < g^{-1}h >)$ inside the RAAG A_Λ so that gA_Λ, hA_Λ are copies of one of the factors. Taking parallelism classes ensures that we will not do multiple counting in the distance formula. What we mean is that infinitely many parallel cosets would give the same contribution to the distance formula, which would clearly break it.

As in the case of $\mathcal{C}S$, $\mathcal{C}(gA_\Lambda)$ is a copy of the Cayley graph of A_Λ with respect to the generating set $V\Lambda \cup \{A_{\Lambda'} < A_\Lambda : \Lambda'$ proper subgraph of $\Lambda\}$. The projection map from A_Γ to $\mathcal{C}(gA_\Lambda)$ is the composition of the closest-point projection to gA_Λ in the usual Cayley graph of A_Γ, and the inclusion $gA_\Lambda \subseteq \mathcal{C}(gA_\Lambda)$. The closest-point projection can also be rephrased in terms of the normal form for elements of A_Γ, since the normal form gives geodesics.

Nesting is inclusion up to parallelism, meaning that we declare $[gA_\Lambda] \sqsubseteq [gA_{\Lambda'}]$ when $\Lambda \subseteq \Lambda'$, where $[\cdot]$ denotes the parallelism class. Similarly, we declare $[gA_\Lambda] \perp [gA_{\Lambda'}]$ if Λ, Λ' form a join.

In the case of RAAGs, it turns out that geodesics in (the usual Cayley graph of) A_Γ are actually hierarchy paths.

Mapping class groups. In this case, $\mathfrak{S}$ is the collection of (isotopy classes of essential) subsurfaces, with each $\mathcal{C}Y$ being the corresponding curve complex, and the maps π_Y are defined using the so-called subsurface projections. Nesting is containment (up to isotopy), while orthogonality corresponds to disjointness (again up to isotopy).

CAT(0) cube complexes. Consider a CAT(0) cube complex $\mathcal{X}$ with a factor system $\mathcal{F}$. In this case, $\mathfrak{S}$ is the union of $\{S = \mathcal{X}\}$ and the set of *parallelism classes* in $\mathcal{F}$. Parallelism can be defined in at least two equivalent ways. The first one is that the convex subcomplexes F, F' are parallel if they cross the same hyperplanes. The second one, which provides a much better picture, is that F, F' are parallel if there exists an isometric embedding of $F \times [0,n] \to \mathcal{X}$, where $[0,n]$ is cubulated by unit intervals and $F \times [0,n]$ is regarded as a cube complex, so that $F \times \{0\}$ maps to F in the obvious way, and the image of $F \times \{n\}$ is F'. As in the case of RAAGs, if we did not take parallelism classes then the distance formula would certainly not work due to multiple counting. We denote parallelism classes by $[F]$.

The $\mathcal{C}[F]$ are obtained starting from F and coning off all $F' \in \mathcal{F}$ contained in F. The maps $\pi_{[F]}$ are defined using gates.

Nesting $[F] \sqsubseteq [F']$ is inclusion up to parallelism, which can also be rephrased as: all hyperplanes crossing F also cross F' (notice that this does not depend on the choice of representatives). Orthogonality $[F] \perp [F']$ means that, up to parallelism, $F \times F'$ has a natural embedding into $\mathcal{X}$. It can also be rephrased as: each hyperplane crossing F crosses each hyperplane crossing F'.

5.2 Technical discussion

Keeping in mind the heuristic discussion from Part 5.1, we now analyse in more detail the definition and the main tools to study HHSs. We start with the axioms.

We will often motivate the axioms in terms of standard product regions, but we warn the reader in advance that those will be constructed only after we discuss all the axioms and a few tools. This, however, is inevitable. In fact, we are trying to describe a space that has some sort of subspaces, the standard product regions, that can be endowed with the same structure as the space itself. Until we know what that structure is in detail, we cannot use it to construct the standard product regions starting from first principles. Hopefully, one or more of the examples we discussed in Part 5.1 can help with intuition.

5.2.1 Commentary on the axioms

We will work in the context of a *quasigeodesic space*, $\mathcal{X}$, i.e., a metric space where any two points can be connected by a uniform-quality quasigeodesic. It is more convenient for us to work with quasi-geodesic metric spaces than geodesic metric spaces because the standard product regions are in a natural way quasi-geodesic metric spaces, rather than geodesic metric spaces. Any quasi-geodesic metric space is quasi-isometric to a geodesic metric space since one can consider an approximating graph whose vertices form a maximal net, so for the purposes of large-scale geometry there's basically no difference between geodesic and quasi-geodesic metric spaces.

Actually, all the requirements in the definition of HHSs are meant to be stable under passing to standard product regions. We do not have standard product regions yet, so what happens in the definition instead is that the axioms are about certain sub-collections of the set of hyperbolic spaces involved in the HHS structure, rather than just the whole collection.

We now go through the definition of HHS given in [BHS1], which is the one with "optimised" axioms compared to [BHS2]. The statements of the axioms are given exactly as in [BHS1].

The q–quasigeodesic space $(\mathcal{X}, \mathsf{d}_{\mathcal{X}})$ *is a* hierarchically hyperbolic space *if there exists $\delta \geq 0$, an index set $\mathfrak{S}$, and a set $\{\mathcal{C}W : W \in \mathfrak{S}\}$ of δ–hyperbolic spaces $(\mathcal{C}U, \mathsf{d}_U)$, such that the following conditions are satisfied:*

1 ***(Projections.)*** *There is a set $\{\pi_W : \mathcal{X} \to 2^{\mathcal{C}W} \mid W \in \mathfrak{S}\}$ of* projections *sending points in $\mathcal{X}$ to sets of diameter bounded by some $\xi \geq 0$ in the various $\mathcal{C}W \in \mathfrak{S}$. Moreover, there exists K so that each π_W is (K,K)–coarsely-Lipschitz and $\pi_W(\mathcal{X})$ is K–quasiconvex in $\mathcal{C}W$.*

The index set $\mathfrak{S}$ is the set of factors of standard product regions. Any $V \in \mathfrak{S}$ hence corresponds to each of many "parallel" subsets of $\mathcal{X}$. We already saw where the hyperbolic spaces associated to an HHS comes from: each factor of a standard product region contains various standard product sub-regions, which we can cone-off to obtain a hyperbolic space. The way to think about the projection is that the standard product regions and their factors come with a coarse retraction from $\mathcal{X}$, and the projections π_W in the definition are the composition of those retractions with the cone-off map. This is admittedly a bit circular because we will later define the retractions in terms of the π_W, but should hopefully help to understand the picture.

The reason that the projections take values in bounded subsets of the $\mathcal{C}W$ rather than points is just that in several situations, for example subsurface projections for mapping class groups, this is what one gets in a natural way. One can make arbitrary choices and modify the projections to take value in points, and nothing would be affected.

2 ***(Nesting.)*** *$\mathfrak{S}$ is equipped with a partial order $\sqsubseteq$, and either $\mathfrak{S} = \emptyset$ or $\mathfrak{S}$ contains a unique $\sqsubseteq$–maximal element; when $V \sqsubseteq W$, we say V is* nested *in W. We require that $W \sqsubseteq W$ for all $W \in \mathfrak{S}$. For each $W \in \mathfrak{S}$, we denote by $\mathfrak{S}_W$ the set of $V \in \mathfrak{S}$ such that $V \sqsubseteq W$. Moreover, for all $V, W \in \mathfrak{S}$ with $V \sqsubsetneq W$ there is a specified subset $\rho^V_W \subset \mathcal{C}W$ with $\mathsf{diam}_{\mathcal{C}W}(\rho^V_W) \leq \xi$. There is also a* projection *$\rho^W_V : \mathcal{C}W \to 2^{\mathcal{C}V}$. (The* *similarity in notation is justified by viewing ρ^V_W as a coarsely constant map $\mathcal{C}V \to 2^{\mathcal{C}W}$.)*

Nesting corresponds to inclusion between standard product regions. The maximal element corresponds to $\mathcal{X}$ itself, thought of as a product region with a trivial factor.

Recall that the $\mathcal{C}W$ are obtained by coning-off standard product regions, i.e. making them bounded. For $V \sqsubset W$, the bounded set ρ^V_W is one such bounded set, where V is regarded as a standard product region with one trivial factor. In the other direction, ρ^W_V is obtained by restricting the retraction to W.

We will discuss below the fact that ρ^W_V for $V \sqsubset W$ is not strictly needed, and can for all intents and purposes be reconstructed from π_W and π_V.

Regarding the notation, the ρ^V_Ws in this axiom as well as the ones below always go "from top to bottom", meaning that ρ^V_W is always some kind of projection from $\mathcal{C}V$ to $\mathcal{C}W$.

3 ***(Orthogonality.)*** *$\mathfrak{S}$ has a symmetric and anti-reflexive relation called* orthogonality*: we write $V \perp W$ when V, W are orthogonal. Also, whenever $V \sqsubseteq W$ and $W \perp U$, we require that $V \perp U$. Finally, we require that for each $T \in \mathfrak{S}$ and each $U \in \mathfrak{S}_T$ for which $\{V \in \mathfrak{S}_T : V \perp U\} \neq \emptyset$, there exists $W \in \mathfrak{S}_T - \{T\}$, so that whenever $V \perp U$ and $V \sqsubseteq T$, we have $V \sqsubseteq W$. Finally, if $V \perp W$, then V, W are not $\sqsubseteq$-comparable.*

Orthogonality is what creates non-trivial products: V and W are orthogonal if they participate in a common standard product region, meaning that they are distinct factors. With this interpretation, it should be clear why when $V \sqsubseteq W$ and $W \perp U$, we require $V \perp U$, and also why when $V \perp W$, then V, W should not be $\sqsubseteq$-comparable.

The tricky part is the one about $\{V \in \mathfrak{S}_T : V \perp U\}$. Let us first discuss the case of $\mathfrak{S}$ instead of more general $\mathfrak{S}_T$. The point is that one wants to define an orthogonal complement of the $V \in \mathfrak{S}$, and one wants it to be an HHS, with corresponding index set $U^{\perp} = \{U \in \mathfrak{S} : V \perp U\}$. For that to be the case, one would want $U^{\perp}$ to contain a $\sqsubseteq$-maximal element (if it is non-empty). The axiom says something a bit weaker, because the W containing each $V \perp U$ is not required to be itself orthogonal to U. This is still enough to have an HHS structure on the orthogonal complement.

The only reason we did not require the stronger version with $W \perp U$ in [BHS2] is that at the time we were not able to prove that such a W exists in the case of CAT(0) cube complexes. However, [HS, Theorem C] says that proper cocompact cube complexes with a factor system satisfy the stronger version of the axiom, so that in fact all natural examples of HHS (so far) do, and there is no harm in strengthening the orthogonality axiom. In fact, sometimes the weaker formulation gives technical problems.

As a final comment, it is natural to formulate the axiom for general $\mathfrak{S}_T$ instead of just for $\mathfrak{S}$ because all axioms need to work inductively for product regions.

4 ***(Transversality and consistency.)*** *If $V, W \in \mathfrak{S}$ are not orthogonal and neither is nested in the other, then we say V, W are* transverse, *denoted $V \pitchfork W$. There exists $\kappa_0 \geq 0$ such that if $V \pitchfork W$, then there are sets $\rho^V_W \subseteq \mathcal{C}W$ and $\rho^W_V \subseteq \mathcal{C}V$ each of diameter at most ξ and satisfying:*

$$\min\left\{\mathsf{d}_W(\pi_W(x), \rho^V_W), \mathsf{d}_V(\pi_V(x), \rho^W_V)\right\} \leq \kappa_0$$

for all $x \in \mathcal{X}$.

For $V, W \in \mathfrak{S}$ satisfying $V \sqsubseteq W$ and for all $x \in \mathcal{X}$, we have:

$$\min\left\{\mathsf{d}_W(\pi_W(x), \rho^V_W), \mathsf{diam}_{\mathcal{C}V}(\pi_V(x) \cup \rho^W_V(\pi_W(x)))\right\} \leq \kappa_0.$$

The preceding two inequalities are the consistency inequalities *for points in $\mathcal{X}$.*

Finally, if $U \sqsubseteq V$, then $\mathsf{d}_W(\rho^U_W, \rho^V_W) \leq \kappa_0$ whenever $W \in \mathfrak{S}$ satisfies either $V \sqsubsetneq W$ or $V \pitchfork W$ and $W \not\perp U$.

Transversality is best thought of as being in "general position". As an aside for the reader who speaks relative hyperbolicity, if $U \pitchfork V$, then they behave very similarly to distinct cosets of peripheral subgroups of a relatively hyperbolic group; for example the projections to $\mathcal{C}U, \mathcal{C}V$ should be compared to closest-point projections onto a pair of distinct cosets. The first consistency inequality, also known as the Behrstock inequality, is very important, so we now discuss a few ways to think about it (and its consequences). Incidentally, we note that the Behrstock inequality is important beyond the HHS world too; for example, it plays a prominent role in the context of the projection complexes from [BBF], which have many applications.

In words, the Behrstock inequality says that if $V \pitchfork W$ and $x \in \mathcal{X}$ projects far from ρ^W_V in $\mathcal{C}V$, then x projects close to ρ^V_W in $\mathcal{C}W$. (ρ^W_V

is best thought of as the projection of W onto $\mathcal{C}V$.) Let us start by discussing an easy situation where the inequality holds. Suppose that V and W are two quasi-convex subsets of a hyperbolic space and suppose that $\pi_V(W), \pi_W(V)$ are both bounded, where π_V, π_W denote (coarse) closest point projections. Then setting $\rho^V_W = \pi_V(W)$ and $\rho^W_V = \pi_W(V)$, the first consistency inequality holds, and it is illustrated in the following picture:

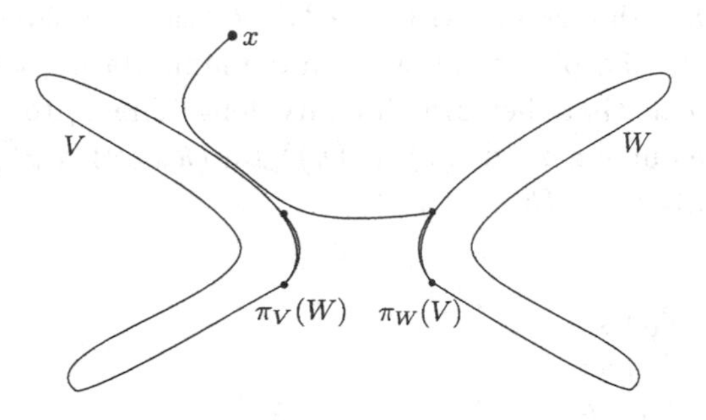

Figure 5.1 The Behrstock inequality for quasiconvex subspaces of a hyperbolic space.

Here is a sketch of the argument, which should also clarify the meaning of the inequality. If x projects to V far away from the projection of W, as in the picture, then we have to show that it projects close to the projection of V onto W. This is because any geodesic from x to W must pass close to V, by a standard hyperbolicity argument.

This last fact is useful to keep in mind: in the situation above, to go from x to W one has to pass close to V first, and change the projection to V in the process.

A second way to understand the inequality is to draw the image of $\pi = \pi_V \times \pi_W$, which is coarsely the following “cross”:

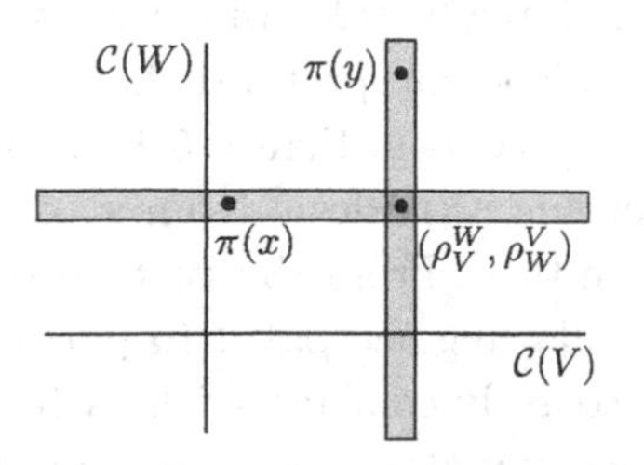

Figure 5.2 The image of $\pi = \pi_V \times \pi_W$, when $V \pitchfork W$.

From this graph we see a similar phenomenon to the one above: depending on where $\pi(x), \pi(y)$ lie on the cross, to go from x to y one has to change the projection to $\mathcal{C}V$ first or the projection to $\mathcal{C}W$ first.

This brings us to an important consequence of the Behrstock inequality, which is that one can order transverse V, W that lie "between" x and y. Suppose that $x, y \in \mathcal{X}$ and $\{V_i\}$ are pairwise transverse and so that $\mathsf{d}_{V_i}(x, y)$ are all much larger than the constant in the Behrstock inequality. Then for each $V, W \in \{V_i\}$, up to switching V, W, the situation looks like Figure 5.2, and in this case we write $V \prec W$. We can give several equivalent descriptions of "the picture above", and manipulating the Behrstock inequality reveals that they are all equivalent. These are the following, where we are assuming $\mathsf{d}_V(\pi_V(x), \pi_V(y)), \mathsf{d}_W(\pi_W(x), \pi_W(y)) \geq 10E$ for some sufficiently large E:

- $V \prec W$,
- $\mathsf{d}_W(\pi_W(x), \rho^V_W) \leq E$,
- $\mathsf{d}_W(\pi_W(y), \rho^V_W) > E$,
- $\mathsf{d}_V(\pi_V(x), \rho^W_V) > E$,
- $\mathsf{d}_V(\pi_V(y), \rho^W_V) \leq E$.

A very important fact is that $\prec$ is a total order on $\{V_i\}$. My favourite way to draw this is the following, assuming for simplicity $V_i \prec V_j$ if and only if $i < j$:

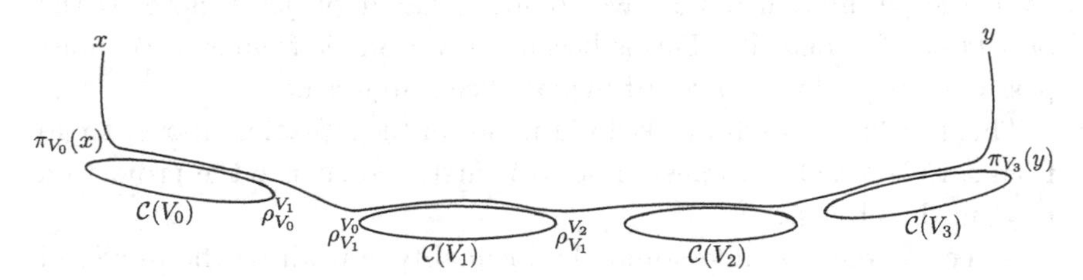

This picture does not really take place anywhere, but it contains interesting information. You can pretend that the $\mathcal{C}V_i$ are quasiconvex subsets of a hyperbolic space as in Figure 5.1, with the path from x to y in the picture representing a geodesic from x to y that passes close to them in the order given by $\prec$. From the picture you can read off where the various ρs are by following the path. In particular, you see that for $i < j < k$, $\rho^{V_j}_{V_i}$ and $\rho^{V_k}_{V_i}$ coarsely coincide with each other and with $\pi_V(y)$. This picture still works to understand where projections lie if you, for example, add another point z. You can try to convince yourself, first from the picture and then formally, that if z projects "in the middle" on some $\mathcal{C}V_i$ then, for $j > i$, $\pi_{V_j}(z)$ coarsely coincides with $\pi_{V_j}(x)$.

We now discuss the second consistency inequality in conjunction with another axiom:

7 **(Bounded geodesic image.)** *For all $W \in \mathfrak{S}$, all $V \in \mathfrak{S}_W - \{W\}$, and all geodesics γ of $\mathcal{C}W$, either $\mathrm{diam}_{\mathcal{C}V}(\rho^W_V(\gamma)) \le E$ or $\gamma \cap \mathcal{N}_E(\rho^V_W) \neq \emptyset$.*

In words: when V is properly nested into W, then the projection ρ^W_V from $\mathcal{C}W$ to $\mathcal{C}V$ is coarsely constant along geodesics far from ρ^V_W (recall that this is the copy of V that gets coned-off to make $\mathcal{C}W$ out of W).

This is virtually always used together with the second consistency inequality, which implies that if $\pi_W(x)$ is far from ρ^V_W for some $x \in \mathcal{X}$, then $\rho^W_V(\pi_W(x))$ coarsely coincides with $\pi_V(x)$. This yields the version of bounded geodesic image that most often gets used in practice:

Lemma 5.2.1 (See e.g. [BHS4, Lemma 1.5]) *Let $(X, \mathfrak{S})$ be hierarchically hyperbolic. Up to increasing E as in the bounded geodesic image axiom, for all $W \in \mathfrak{S}$, all $V \in \mathfrak{S}_W - \{W\}$, and all $x, y \in \mathcal{X}$ so that some geodesic from $\pi_W(x)$ to $\pi_W(y)$ stays E–far from ρ^V_W, we have $\mathrm{d}_V(\pi_V(x), \pi_V(y)) \le E$.*

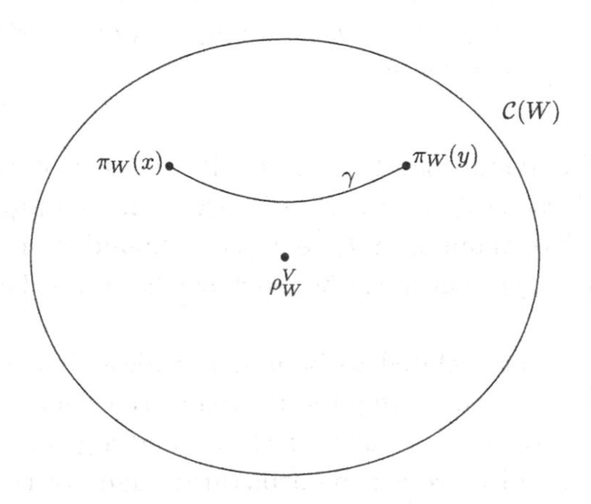

Figure 5.3 In the picture we have $V \sqsubseteq W$ and γ is a geodesic. According to bounded geodesic image, $\pi_V(x)$ and $\pi_V(y)$ coarsely coincide.

One can simply replace the bounded geodesic image axiom and the second consistency inequality with the lemma, since ρ^W_V can be reconstructed from π_W and π_V at least on $\pi_W(\mathcal{X})$ in view of the lemma. However, for some purposes one still needs ρ^W_V. This is most notably the case for the realisation theorem.

Another picture to keep in mind regarding bounded geodesic image is that, given $x, y \in \mathcal{X}$ and W, one can consider all $V \sqsubset W$ so that $\mathsf{d}_V(\pi_V(x), \pi_V(y))$ is large. The corresponding ρ^V_W will form a "halo" around a geodesic from $\pi_W(x)$ to $\pi_W(y)$.

5 ***(Finite complexity.)*** *There exists $n \geq 0$, the* complexity *of $\mathcal{X}$ (with respect to $\mathfrak{S}$), so that any set of pairwise-$\sqsubseteq$-comparable elements has cardinality at most n.*

This axiom should be pretty self-explanatory. Induction on complexity is very common in the HHS world. The base case (complexity 1) is that of hyperbolic spaces.

6 ***(Large links.)*** *There exists $E \geq \max\{\xi, \kappa_0\}$ such that the following holds. Let $W \in \mathfrak{S}$ and let $x, x' \in \mathcal{X}$. Let $N = \mathsf{d}_W(\pi_W(x), \pi_W(x'))$. Then there exist $T_1, \ldots, T_{\lfloor N \rfloor} \in \mathfrak{S}_W - \{W\}$ such that for all $T \in \mathfrak{S}_W - \{W\}$, either $T \in \mathfrak{S}_{T_i}$ for some i, or $\mathsf{d}_T(\pi_T(x), \pi_T(x')) < E$. Also, $\mathsf{d}_W(\pi_W(x), \rho^{T_i}_W) \leq N$ for each i.*

In words, the axioms say that, given W and $x, x' \in \mathcal{X}$, each of the $V \sqsubset W$ such that $\mathsf{d}_V(\pi_V(x), \pi_V(x'))$ is large is nested into one of a few fixed $T_i \sqsubset W$. The number of T_i required is bounded only in terms of $\mathsf{d}_W(\pi_W(x), \pi_W(x'))$ (which can be much smaller than their distance in $\mathcal{X}$).

This axiom is very related to bounded geodesic image, and in fact in concrete examples they are often proven at the same time. Bounded geodesic image provides a "halo" of ρ^V_W around a geodesic connecting $\pi_W(x), \pi_W(x')$, and there can be arbitrarily many of these. However, large links organises them into a few (possibly intersecting) subsets, each of which contains the ρ^V_W with V nested into some fixed T_i. The number of such T_i is bounded in terms of the distance $\mathsf{d}_W(\pi_W(x), \pi_W(x'))$.

Large links is used in arguments of the following type. Consider two points x, y that are far in $\mathcal{X}$. If they are far in $\mathcal{C}S$ (meaning that their projections are), then one can use the geometry of $\mathcal{C}S$ to study whatever property one is interested in. Otherwise, there are few T_is, and one can then analyse corresponding standard product regions. In one of them, (the retractions of) x, y are still far away, so one can use induction based on the fact that the standard product region is an HHS of strictly lower complexity.

One concrete lemma that makes this more precise is the "passing up" lemma [BHS1, Lemma 2.5]. This says the (contrapositive of the) following. If one has $x, y \in \mathcal{X}$ and some $S_i \in \mathfrak{S}$ so that the $\mathsf{d}_{S_i}(\pi_{S_i}(x), \pi_{S_i}(y))$ are all large and each S_i is $\sqsubseteq$–maximal with this property, then there is a bound on how many S_i there are.

8 ***(Partial realization.)*** *There exists a constant α with the following property. Let $\{V_j\}$ be a family of pairwise orthogonal elements of $\mathfrak{S}$, and let $p_j \in \pi_{V_j}(\mathcal{X}) \subseteq \mathcal{C}V_j$. Then there exists $x \in \mathcal{X}$ so that:*

- $\mathsf{d}_{V_j}(x, p_j) \leq \alpha$ *for all* j,
- *for each j and each $V \in \mathfrak{S}$ with $V_j \sqsubset V$, we have $\mathsf{d}_V(x, \rho^{V_j}_V) \leq \alpha$, and*
- *if $W \pitchfork V_j$ for some j, then $\mathsf{d}_W(x, \rho^{V_j}_W) \leq \alpha$.*

Roughly speaking, the axiom says that, given pairwise-orthogonal $\{V_i\}$ there is no restriction on the projections of points of $\mathcal{X}$ to the $\mathcal{C}V_i$; any choice of coordinates can be realised by a point in $\mathcal{X}$. This is the opposite of what happens when $V \not\perp W$ (i.e. in one of the cases $V = W, V \sqsubset W, W \sqsubset V$ or $V \pitchfork W$), in which case there are serious restrictions on the projections in view of the consistency inequalities.

This axiom gives us the first glimpse of how the standard product regions arise, and what their coordinates in the various $\mathcal{C}U$ look like. Starting from the family of pairwise orthogonal elements $\{V_j\}$, we see that the axioms provides us with the freedom to move independently in each of the $\mathcal{C}V_j$. When we will have the "full" realisation theorem, this will give us a product region associated to $\{V_j\}$. The second condition can be explained as follows: the coordinate in $\mathcal{C}V$ does not coarsely vary when moving around the standard product region because the standard product region is coned-off there. The third condition tells us that "generic" pairs of standard product regions do not interact much with each other. (Recall that we think of transversality as being in "generic position".)

9 ***(Uniqueness.)*** *For each $\kappa \geq 0$, there exists $\theta_u = \theta_u(\kappa)$ such that if $x, y \in \mathcal{X}$ and $\mathsf{d}(x, y) \geq \theta_u$, then there exists $V \in \mathfrak{S}$ such that $\mathsf{d}_V(x, y) \geq \kappa$.*

Informally, the axiom says that if x, y are close in each $\mathcal{C}V$ (meaning that their projections are), then x, y are close in $\mathcal{X}$.

This axiom is a weaker form of the distance formula. The point is that it is in many circumstances much easier to prove than the "full" distance formula, allowing for easier proofs that certain spaces are HHS. This is the case for mapping class groups, where there's a one-page argument for this axiom, given in [BHS1, Section 11], while the known proofs of the distance formula are much more involved.

5.2.2 Main tools

In addition to the axioms, there are 3 fundamental properties of HHSs. These were actually part of the first set of axioms, but they have a much higher level of sophistication than any of the axioms.

Distance formula

We stated the distance formula in Part 5.1, but let us recall it. Given $A, B \in \mathbb{R}$, the symbol $\{\!\{A\}\!\}_B$ will denote A if $A \geq B$ and 0 otherwise. Given C, D, we write $A \asymp_{C,D} B$ to mean $C^{-1}A - D \leq B \leq CA + D$.

To save notation, we denote $\mathsf{d}_W(x, y) = \mathsf{d}_W(\pi_W(x), \pi_W(y))$.

Theorem 5.2.2 (Distance Formula, [BHS1, Theorem 4.5]) *Let $(X, \mathfrak{S})$ be hierarchically hyperbolic. Then there exists s_0 such that for all $s \geq s_0$ there exist constants K, C such that for all $x, y \in \mathcal{X}$,*

$$\mathsf{d}_{\mathcal{X}}(x, y) \asymp_{K,C} \sum_{W \in \mathfrak{S}} \{\!\{\mathsf{d}_W(x, y)\}\!\}_s .$$

The distance formula allows one to reconstruct the geometry of $\mathcal{X}$ from that of the hyperbolic spaces $\mathcal{C}W$, and at this point its importance should hopefully be evident. It is important to note that the distance formula works for any sufficiently high threshold. This is useful in practice because typically one proceeds along the following lines. One starts with a configuration in $\mathcal{X}$, projects it to the $\mathcal{C}W$ and keeps in account the distance formula to figure out what one gets. Then one performs some coarse construction in the $\mathcal{C}W$, and then goes back to $\mathcal{X}$. In the process, more often than not some projections get moved a bounded amount. To compensate for this, one uses a higher threshold in the distance formula.

Hierarchy paths

Hierarchy paths are quasi-geodesics in $\mathcal{X}$ that shadow geodesics in all $\mathcal{C}W$, which is clearly a very nice property to have since we want to relate the geometry of $\mathcal{X}$ to that of the $\mathcal{C}W$. Let us define them precisely.

For M a metric space, a (coarse) map $f\colon [0,\ell] \to M$ is a (D,D)–*unparametrised quasigeodesic* if there exists a strictly increasing function $g\colon [0,L] \to [0,\ell]$ such that $f \circ g\colon [0,L] \to M$ is a (D,D)–quasigeodesic and for each $j \in [0,L] \cap \mathbb{N}$, we have $\mathsf{diam}_M\left(f(g(j)) \cup f(g(j+1))\right) \le D$.

Definition 5.2.3 (Hierarchy path) Let $(X, \mathfrak{S})$ be hierarchically hyperbolic. For $D \ge 1$, a (not necessarily continuous) path $\gamma\colon [0,\ell] \to \mathcal{X}$ is a *D–hierarchy path* if

1 γ is a (D,D)-quasigeodesic,
2 for each $W \in \mathfrak{S}$, the path $\pi_W \circ \gamma$ is an unparametrised (D,D)–quasigeodesic.

Theorem 5.2.4 (Existence of Hierarchy Paths, [BHS1, Theorem 4.4]) *Let $(\mathcal{X}, \mathfrak{S})$ be hierarchically hyperbolic. Then there exists D_0 so that any $x, y \in \mathcal{X}$ are joined by a D_0-hierarchy path.*

Whenever possible, one should work with hierarchy paths rather than other quasi-geodesics, even actual geodesics. Unfortunately, not all quasi-geodesics are hierarchy paths (meaning that one cannot control how close the projection to some $\mathcal{C}W$ of a (D,D)–quasigeodesic is to being a geodesic as a function of D only). In fact, there are spiralling quasi-geodesics in $\mathbb{R}^2$, and, even worse than that, it is a folklore result that in mapping class groups there are quasi-geodesics that project to "arbitrarily bad" paths even in the curve graph of the whole surface.

Moreover, hierarchy paths with given endpoints are not coarsely unique: think of $\mathbb{R}^2$, where there are plenty of quasi-geodesics monotone in each factor that connect points far away along a diagonal. In fact, it is a very important problem to study to what extent one can make hierarchy paths canonical by adding more restrictions.

Realisation

In this subsection we discuss the realisation theorem, which says that the consistency inequalities characterise the tuples $(\pi_W(x))_{W \in \mathfrak{S}}$ for $x \in \mathcal{X}$. We think of the $\pi_W(x)$ as the coordinates of x.

Definition 5.2.5 (Consistent) Fix $\kappa \ge 0$ and let $\vec{b} \in \prod_{U \in \mathfrak{S}} 2^{\mathcal{C}U}$ be a tuple such that for each $U \in \mathfrak{S}$, the coordinate b_U is a subset of $\mathcal{C}U$ with $\mathsf{diam}_{\mathcal{C}U}(b_U) \le \kappa$. The tuple $\vec{b}$ is *κ–consistent* if $\mathsf{d}_U(b_U, \pi_U(\mathcal{X})) \le \kappa$ for all $U \in \mathfrak{S}$, and the following holds. Whenever $V \pitchfork W$,

$$\min\left\{\mathsf{d}_W(b_W, \rho_W^V), \mathsf{d}_V(b_V, \rho_V^W)\right\} \le \kappa,$$

and whenever $V \sqsubseteq W$,

$$\min\{\mathsf{d}_W(b_W, \rho_W^V), \mathsf{diam}_{CV}(b_V \cup \rho_V^W(b_W))\} \leq \kappa.$$

(Notice that in the definition of a consistent tuple we need the map ρ_V^W for $V \sqsubset W$.)

Theorem 5.2.6 (Realisation of consistent tuples, [BHS1, Theorem 3.1]) *For each $\kappa \geq 1$ there exist $\theta_e, \theta_u \geq 0$ such that the following holds. Let $\vec{b} \in \prod_{W \in \mathfrak{S}} 2^{CW}$ be κ–consistent; for each W, let b_W denote the CW–coordinate of $\vec{b}$.*

Then there exists $x \in \mathcal{X}$ so that $\mathsf{d}_W(b_W, \pi_W(x)) \leq \theta_e$ for all $CW \in \mathfrak{S}$. Moreover, x is coarsely unique *in the sense that the set of all x that satisfy $\mathsf{d}_W(b_W, \pi_W(x)) \leq \theta_e$ in each $CW \in \mathfrak{S}$, has diameter at most θ_u.*

As mentioned in Part 5.1, the realisation theorem is used to perform constructions in all the CW separately and then pull those back to $\mathcal{X}$. One such construction is (at last!) that of standard product regions. Basically, we fix $U \in \mathfrak{S}$, and consider partial systems of coordinates (b_V), where we only assign b_V when either $V \sqsubseteq U$ or $V \perp U$. If this partial system of coordinates satisfies the consistency inequalities, we can extend it and use realisation to find a corresponding point in $\mathcal{X}$. The standard product region associated to U is obtained considering all such realisation points. A similar game can be played starting from pairwise orthogonal U_i, but for simplicity we stick to the case of a single U. Let us make this more precise.

Definition 5.2.7 (Nested partial tuple) Recall $\mathfrak{S}_U = \{V \in \mathfrak{S} : V \sqsubseteq U\}$. Fix $\kappa \geq \kappa_0$ and let $\mathbf{F}_U$ be the set of κ–consistent tuples in $\prod_{V \in \mathfrak{S}_U} 2^{CV}$.

Definition 5.2.8 (Orthogonal partial tuple) Let $\mathfrak{S}_U^{\perp} = \{V \in \mathfrak{S} : V \perp U\} \cup \{A\}$, where A is a $\sqsubseteq$–minimal element W such that $V \sqsubseteq W$ for all $V \perp U$. Fix $\kappa \geq \kappa_0$, let $\mathbf{E}_U$ be the set of κ–consistent tuples in $\prod_{V \in \mathfrak{S}_U^{\perp} - \{A\}} 2^{CV}$.

Construction 5.2.9 (Product regions in $\mathcal{X}$) Given $\mathcal{X}$ and $U \in \mathfrak{S}$, there are coarsely well-defined maps $\phi^{\sqsubseteq}, \phi^{\perp} : \mathbf{F}_U, \mathbf{E}_U \to \mathcal{X}$ which extend to a coarsely well-defined map $\phi_U : \mathbf{F}_U \times \mathbf{E}_U \to \mathcal{X}$, whose image P_U we call a *standard product region.* Indeed, for each $(\vec{a}, \vec{b}) \in \mathbf{F}_U \times \mathbf{E}_U$, and each $V \in \mathfrak{S}$, define the co-ordinate $(\phi_U(\vec{a}, \vec{b}))_V$ as follows. If $V \sqsubseteq U$, then $(\phi_U(\vec{a}, \vec{b}))_V = a_V$. If $V \perp U$, then $(\phi_U(\vec{a}, \vec{b}))_V = b_V$. If $V \pitchfork U$, then $(\phi_U(\vec{a}, \vec{b}))_V = \rho_V^U$. Finally, if $U \sqsubseteq V$, and $U \neq V$, let $(\phi_U(\vec{a}, \vec{b}))_V = \rho_V^U$.

By design of the axioms, it is straightforward (but a bit tedious) to check that we actually defined a consistent tuple, see [BHS2, Section 13.1].

We notice that the following hold by the very definition of P_U. First, $\pi_Y(P_U)$ is uniformly bounded if $U \sqsubsetneq Y$ (making sure that it makes sense to think of U as being coned-off to get $\mathcal{C}Y$), as well as if $U \pitchfork Y$ (so that we can actually think of P_U and P_Y as "independent").

Coarse retractions onto standard product regions have been mentioned above. It should not be hard to guess how they are constructed at this point. One simply starts with $x \in \mathcal{X}$, defines coordinates by taking $\pi_Y(x)$ whenever $Y \sqsubseteq U$ or $Y \perp U$ and ρ^U_Y otherwise, and takes a realisation point. Basically, one defines the retraction of $x \in \mathcal{X}$ by keeping the coordinates involved in the standard product region only. This is a special case of the gate maps discussed in Subsection 5.2.3 below.

One useful picture to keep in mind is that essentially the only efficient way to move from some $x \in \mathcal{X}$ to some standard product region P_U is to go towards the gate of x in P_U, and then move in P_U. When moving from x to its gate, one only changes the projection to some $\mathcal{C}Y$ if either $Y \pitchfork U$ or $U \sqsubsetneq Y$, since these are the $\mathcal{C}Y$ where x and its gate can have different projections.

Coarse median. As mentioned in Part 5.1, the realisation theorem can be used to construct a coarse-median map in the sense of [Bo1] (also called centroid in [BM]). This is the map $\mathfrak{m}\colon \mathcal{X}^3 \to \mathcal{X}$ defined as follows. Let $x, y, z \in \mathcal{X}$ and, for each $U \in \mathfrak{S}$, let b_U be a coarse centre for the triangle with vertices $\pi_U(x), \pi_U(y), \pi_U(z)$. More precisely, b_U is any point in $\mathcal{C}U$ with the property that there exists a geodesic triangle in $\mathcal{C}U$ with vertices in $\pi_U(x), \pi_U(y), \pi_U(z)$ each of whose sides contains a point within distance δ of b_U, where δ is the hyperbolicity constant of $\mathcal{C}U$.

By [BHS1, Lemma 2.6] (which is easy, and a good exercise), the tuple $\vec{b} \in \prod_{U \in \mathfrak{S}} 2^{\mathcal{C}U}$ whose U-coordinate is b_U is κ–consistent for an appropriate choice of κ. Hence, by the realisation theorem, there exists $\mathfrak{m} = \mathfrak{m}(x, y, z) \in \mathcal{X}$ such that $\mathsf{d}_U(\mathfrak{m}, b_U)$ is uniformly bounded for all $U \in \mathfrak{S}$. Moreover, this is coarsely well-defined, by the uniqueness axiom. The fact that this coarse median map actually makes the HHS into a coarse median space, and that, moreover, the rank is the "expected" one, is [BHS4, Corollary 2.15].

The existence of the coarse median has many useful consequences, for example regarding asymptotic cones [Bo1, Bo2] ([BHS4] heavily relies on these). Also, the explicit construction itself is useful in various arguments, for example to construct the kind of retractions mentioned in the subsection on hierarchical quasiconvexity below.

5.2.3 Additional tools

Hierarchical quasiconvexity

Quasiconvex subspaces are important in the study of hyperbolic spaces. The corresponding notion in the HHS world is *hierarchical quasiconvexity*. Prominent examples of hierarchically quasiconvex subspaces are standard product regions.

The natural first guess for what a hierarchically quasiconvex subspace of an HHS $\mathcal{X}$ should be is a subspace that projects to uniformly quasiconvex subspaces in all $\mathcal{C}U$. This is part of the definition, but not quite enough to have a good notion. In fact, the aforementioned property is satisfied by subspaces that are not even coarsely connected. There are at least two ideas to "complete" the definition.

The first idea, and the one leading to the definition given in [BHS1], is that not only the projections to the $\mathcal{C}U$ should be quasiconvex, but they should also determine the subspace. One ensures that this is the case by requiring that all realisation points of coordinates $(b_U \in \pi_U(Q))$ lie close to Q, where Q is the hierarchically quasiconvex subspace. That also ensures that Q is an HHS itself, see [BHS1, Proposition 5.6].

A picture that one should keep in mind regarding this second property is that it is not satisfied by an "L" in $\mathbb{R}^2$, even though its projections to the two factors are convex. Rather, what one wants is a "full" square. This is related to the difference, in the cubical world, between ℓ_1–isometric embeddings (the "L") as opposed to convex embeddings (the square).

The second idea to complete the definition is probably more intuitive. Unfortunately, as of yet it has not been proven that this gives an equivalent notion. Recall that given an HHS, there is a preferred family of quasi-geodesics connecting pairs of points, which are called hierarchy paths, and that their defining property is that they project to unparametrised quasi-geodesics in all $\mathcal{C}U$. It is then natural to just replace geodesics in the definition of the usual quasiconvexity in hyperbolic spaces with hierarchy paths. Namely we want to say that Q is hierarchically quasiconvex if all hierarchy paths joining points of Q stay close to Q. It is easy to see that this implies that the projections of Q to the $\mathcal{C}U$ are quasiconvex, but the additional property is not clear, and it would be interesting to known whether it is satisfied. It is however true (and easy to see) that hierarchy paths joining points on a hierarchically quasiconvex set stay close to it. Moreover, the hull of two points, as defined in the next subsection, coarsely coincides with the union of all hierarchy paths (with fixed, large enough, constant) connecting them.

Definition 5.2.10 (Hierarchical quasiconvexity) [BHS1, Definition 5.1] Let $(\mathcal{X}, \mathfrak{S})$ be a hierarchically hyperbolic space. Then $\mathcal{Y} \subseteq \mathcal{X}$ is *k–hierarchically quasiconvex*, for some $k\colon [0,\infty) \to [0,\infty)$, if the following hold:

1 For all $U \in \mathfrak{S}$, the projection $\pi_U(\mathcal{Y})$ is a $k(0)$–quasiconvex subspace of the δ–hyperbolic space $\mathcal{C}U$.
2 For all $\kappa \geq 0$ and κ-consistent tuples $\vec{b} \in \prod_{U \in \mathfrak{S}} 2^{\mathcal{C}U}$ with $b_U \subseteq \pi_U(\mathcal{Y})$ for all $U \in \mathfrak{S}$, each point $x \in \mathcal{X}$ for which $\mathsf{d}_U(\pi_U(x), b_U) \leq \theta_e(\kappa)$ (where $\theta_e(\kappa)$ is as in Theorem 5.2.6) satisfies $\mathsf{d}(x, \mathcal{Y}) \leq k(\kappa)$.

Remark 5.2.11 Note that condition (2) in the above definition is equivalent to: for every $\kappa > 0$, every point $x \in \mathcal{X}$ satisfying

$$\mathsf{d}_U(\pi_U(x), \pi_U(\mathcal{Y})) \leq \kappa$$

for all $U \in \mathfrak{S}$ has the property that $\mathsf{d}(x, \mathcal{Y}) \leq k(\kappa)$.

A very important property of hierarchically quasiconvex subspaces is that they admit a natural coarse retraction, which generalises the retraction onto standard product regions. The definition (see [BHS1, Definition 5.4]) is that a coarsely-Lipschitz map $\mathfrak{g}_{\mathcal{Y}}\colon \mathcal{X} \to \mathcal{Y}$ is called a *gate map* if for each $x \in \mathcal{X}$ it satisfies: $\mathfrak{g}_{\mathcal{Y}}(x)$ is a point $y \in \mathcal{Y}$ such that for all $V \in \mathfrak{S}$, the set $\pi_V(y)$ (uniformly) coarsely coincides with the projection of $\pi_V(x)$ to the $k(0)$–quasiconvex set $\pi_V(\mathcal{Y})$.

The uniqueness axiom implies that when such a map exists it is coarsely well-defined, and the existence is proven in [BHS1, Lemma 5.5]. A useful reference for several properties of gate maps is [BHS4, Lemma 1.19], which provides a lot of information about gate maps between hierarchically quasiconvex subspaces (roughly, if A, B are hierarchically quasiconvex, then $\mathfrak{g}_A(B)$ and $\mathfrak{g}_B(A)$ are hierarchically quasiconvex, "parallel", they each represent the coarse intersection of A and B, and there is a "bridge" of the form $\mathfrak{g}_A(B) \times \mathcal{Y}$ connecting the gates).

Hulls and their cubulation

Other important examples of hierarchically quasiconvex subspaces are hulls. In a δ-hyperbolic space X, if one considers any subset $A \subseteq X$, then one can construct a "quasiconvex hull" simply by taking the union of all geodesics connecting pairs of points in the space. It is easy to show that this is a 2δ-quasiconvex subspace, whatever A is. In an HHS, we can proceed similarly.

Definition 5.2.12 (Hull of a set; [BHS1]) For each $A \subset \mathcal{X}$ and $\theta \geq 0$, let the *hull*, $H_\theta(A)$, be the set of all $p \in \mathcal{X}$ so that, for each $W \in \mathfrak{S}$, the set $\pi_W(p)$ lies at distance at most θ from $\mathrm{hull}_{\mathcal{C}W}(A)$, the convex hull of A in the hyperbolic space $\mathcal{C}W$ (that is to say, the union of all geodesics in $\mathcal{C}W$ joining points of A). Note that $A \subset H_\theta(A)$.

It is proven in [BHS1, Lemma 6.2] that for each sufficiently large θ there exists $\kappa \colon \mathbb{R}_+ \to \mathbb{R}_+$ so that for each $A \subset \mathcal{X}$ the set $H_\theta(A)$ is κ–hierarchically quasiconvex.

It feels like there should be a characterisation of hulls in terms of hierarchy paths. It is not true that the hull of a set coarsely coincides with the union of all hierarchy paths (with suitable constant) connecting pairs of points on the set; there are counterexamples in products of trees. However, it is plausible that one coarsely obtains the hull if one repeats the procedure of taking the union of all hierarchy paths some number of times bounded by a function of the complexity.

To connect a few things we have seen so far, the gate map onto the hull of two points x, y coincides with taking the median $\mathfrak{m}(x, y, \cdot)$.

Hulls of finite sets carry more structure than just being hierarchically quasiconvex. In a hyperbolic space, hulls of finitely many points are quasi-isometric to trees, with constants depending only on how many points one is considering. Trees are 1-dimensional CAT(0) cube complexes, and what happens in more general HHSs is that hulls of finitely many points are quasi-isometric to possibly higher dimensional CAT(0) cube complexes:

Theorem 5.2.13 *[BHS4, Theorem C] Let $(\mathcal{X}, \mathfrak{S})$ be a hierarchically hyperbolic space and let $k \in \mathbb{N}$. Then there exists M_0 so that for all $M \geq M_0$ there is a constant C_1 so that for any $A \subset \mathcal{X}$ of cardinality $\leq k$, there is a C_1–quasimedian[3] (C_1, C_1)–quasi-isometry $\mathfrak{p}_A \colon \mathcal{Y} \to H_\theta(A)$.*

Moreover, let $\mathcal{U}$ be the set of $U \in \mathfrak{S}$ so that $\mathrm{d}_U(x, y) \geq M$ for some $x, y \in A$. Then $\dim \mathcal{Y}$ is equal to the maximum cardinality of a set of pairwise-orthogonal elements of $\mathcal{U}$.

Finally, there exist 0–cubes $y_1, \dots, y_{k'} \in \mathcal{Y}$ so that $k' \leq k$ and $\mathcal{Y}$ is equal to the convex hull in $\mathcal{Y}$ of $\{y_1, \dots, y_{k'}\}$.

The theorem above is crucial for the proof of the quasiflats theorem in [BHS4], and it definitely feels as if it should have many more applications.

[3] CAT(0) cube complexes have a natural median map. Here by C_1–quasimedian we mean that the coarse median of 3 points is mapped within distance C_1 of the median of their images.

Factored spaces

In this section we discuss a construction where one cones-off subspaces of an HHS and obtains a new HHS. Examples of HHSs that are obtained using this construction from some other HHS are the "main" hyperbolic space $\mathcal{C}S$, which is obtained by coning-off standard product regions in the corresponding $\mathcal{X}$, and the Weil–Petersson metric on Teichmüller space, which is (coarsely) obtained by coning-off Dehn twist flats in the corresponding mapping class group. The construction was devised in [BHS3], with the purpose of having spaces that "interpolate" between a given HHS $\mathcal{X}$ and the hyperbolic space $\mathcal{C}S$. In particular, this gives one way of doing induction arguments.

Given a hierarchical space $(\mathcal{X}, \mathfrak{S})$, we say $\mathfrak{U} \subseteq \mathfrak{S}$ is *closed under nesting* if for all $U \in \mathfrak{U}$, if $V \in \mathfrak{S} - \mathfrak{U}$, then $V \not\sqsubseteq U$.

Definition 5.2.14 (Factored space) [BHS3, Definition 2.1] Let $(\mathcal{X}, \mathfrak{S})$ be a hierarchically hyperbolic space. A *factored space* $\widehat{\mathcal{X}}_{\mathfrak{U}}$ is constructed by defining a new metric $\hat{\mathsf{d}}$ on $\mathcal{X}$ depending on a given subset $\mathfrak{U} \subset \mathfrak{S}$ that is closed under nesting. First, for each $U \in \mathfrak{U}$, for each pair $x, y \in \mathcal{X}$ for which there exists $e \in \mathbf{E}_U$ such that $x, y \in \mathbf{F}_U \times \{e\}$, we set $\mathsf{d}'(x,y) = \min\{1, \mathsf{d}(x,y)\}$. For any pair $x, y \in \mathcal{X}$ for which there does not exists such an e we set $\mathsf{d}'(x,y) = \mathsf{d}(x,y)$. We now define the distance $\hat{\mathsf{d}}$ on $\widehat{\mathcal{X}}_{\mathfrak{U}}$. Given a sequence $x_0, x_1, \dots, x_k \in \widehat{\mathcal{X}}_{\mathfrak{U}}$, define its length to be $\sum_{i=1}^{k-1} \mathsf{d}'(x_i, x_{i+1})$. Given $x, x' \in \widehat{\mathcal{X}}_{\mathfrak{U}}$, let $\hat{\mathsf{d}}(x, x')$ be the infimum of the lengths of such sequences $x = x_0, \dots, x_k = x'$.

It is proven in [BHS3, Proposition 2.4] that factored spaces are HHSs themselves, with the "obvious" substructure of the original HHS structure.

I believe that factored spaces will play an important role in studying quasi-isometries between hierarchically hyperbolic spaces in view of [BHS4, Corollary 6.3], which says that a quasi-isometry between HHSs induces a quasi-isometry between certain factored spaces. One can then take full advantage of the HHS machinery to iterate this procedure and get more information about the quasi-isometry under examination. This strategy should work in various sub-classes of right-angled Artin and Coxeter groups (one example is given in [BHS4]).

Boundary

HHSs admit a boundary, defined in [DHS]. Here is the idea behind the definition. We have to understand how one goes to infinity in an HHS, and more precisely we want to understand how one goes to infinity along a hierarchy ray, because we like hierarchy paths. The hierarchy ray will project to any given $\mathcal{C}U$ close to either a geodesic segment or a geodesic ray. In fact, a consequence of large links is (the hopefully intuitive fact) that there must be some $\mathcal{C}U$ so that the hierarchy ray has unbounded projection there (see the useful "passing up" lemma [BHS1, Lemma 2.5]). Now, one can consider the non-empty set of all $\mathcal{C}U$ where the hierarchy ray has unbounded projection, and it should come to no surprise that they must be pairwise orthogonal, because in all other cases there are constraints coming from the consistency inequalities on the projections of points of $\mathcal{X}$. As a last thing, you might want to measure how fast you are moving asymptotically in each of the $\mathcal{C}U$ and record the ratios of the various speeds. These considerations lead to the following definition.

Definition 5.2.15 [DHS, Definitions 2.2-2.3] A *support set* $\overline{S} \subset \mathfrak{S}$ is a set with $S_i \perp S_j$ for all $S_i, S_j \in \overline{S}$. Given a support set $\overline{S}$, a *boundary point* with *support* $\overline{S}$ is a formal sum $p = \sum_{S \in \overline{S}} a_S^p p_S$, where each $p_S \in \partial \mathcal{C}S$, and $a_S^p > 0$, and $\sum_{S \in \overline{S}} a_S^p = 1$. (Such sums are necessarily finite.)

The *HHS boundary* $\partial(\mathcal{X}, \mathfrak{S})$ of $(\mathcal{X}, \mathfrak{S})$ is the set of boundary points.

The boundary also comes with a topology, which is unfortunately very complicated to define. But, fortunately, for at least some applications one can just use a few of its properties and never work with the actual definition (this is the case for the rank rigidity theorems [DHS, Theorems 9.13,914]). The main property is that if an HHS is proper as a metric space, then its boundary is compact by [DHS, Theorem 3.4]. Another good property is that hierarchically quasiconvex subspaces (and not only those) have well-defined boundary extensions.

Modifying the HHS structure

There are various way to modify an HHS structure, and this is useful to perform certain constructions. For example, there is a combination theorem for trees of HHSs which, unsurprisingly, requires the HHS structures of the various edge and vertex groups to be "compatible". In concrete cases, one might be starting with a tree of HHSs that does not satisfy the compatibility conditions on the nose, but does after suitably adjusting the various HHS structures. This is the case even for very simple examples such as the amalgamated product of two mapping class groups over maximal virtually cyclic subgroups containing a pseudo-Anosov.

A typical way of modifying the HHS structure is to cone-off a collection of quasi-convex subspaces of one of the $\mathcal{C}U$, and then add those quasi-convex subspaces as new hyperbolic spaces. This usually amounts to considering certain (hierarchically quasiconvex and hyperbolic) subspaces of $\mathcal{X}$ as product regions with a trivial factor. This construction is used to set things up for the small cancellation constructions in [BHS3], and is explored in much more depth in [S2]. In the latter paper the author studies very general families of quasiconvex subspaces of $\mathcal{C}S$ (called factor systems) so that one can perform a version of the cone-off-and-add-separately construction mentioned above. This is needed to set things up before using the combination theorem in natural examples, like the one mentioned above.

In another direction, one may wonder whether there exists a minimal HHS structure, and in particular one might want to reverse the procedure above. This is explored in [ABD].

References

[ABD] Carolyn Abbott, Jason Behrstock, and Matthew Gentry Durham. Largest acylindrical actions and stability in hierarchically hyperbolic groups. ARXIV:1705.06219, 2017.

[Be] Jason A. Behrstock. Asymptotic geometry of the mapping class group and Teichmüller space. *Geom. Topol.*, 10:1523–1578, 2006.

[BHS1] Jason Behrstock, Mark F Hagen, and Alessandro Sisto. Hierarchically hyperbolic spaces II: combination theorems and the distance formula. ARXIV:1509.00632, 2015.

[BHS2] Jason Behrstock, Mark Hagen, and Alessandro Sisto. Hierarchically hyperbolic spaces, I: Curve complexes for cubical groups. *Geom. Topol.*, 21(3):1731–1804, 2017.

[BHS3] Jason Behrstock, Mark F Hagen, and Alessandro Sisto. Asymptotic dimension and small-cancellation for hierarchically hyperbolic spaces and groups. *Proc. Lond. Math. Soc.*, 114(5):890–926, 2017.

[BHS4] Jason Behrstock, Mark F Hagen, and Alessandro Sisto. Quasiflats in hierarchically hyperbolic spaces. ARXIV:1704.04271, 2017.

[BM] Jason A. Behrstock and Yair N. Minsky. Centroids and the rapid decay property in mapping class groups. *J. Lond. Math. Soc. (2)*, 84(3):765–784, 2011.

[BBF] Mladen Bestvina, Ken Bromberg, and Koji Fujiwara. Constructing group actions on quasi-trees and applications to mapping class groups. *Publ. Math. Inst. Hautes Études Sci.*, 122:1–64, 2015.

[Bo1] Brian H. Bowditch. Coarse median spaces and groups. *Pacific J. Math.*, 261(1):53–93, 2013.

[Bo2] Brian H Bowditch. Large-scale rigidity properties of the mapping class groups. *To appear in Pacific J. Math.*, 2017.

[DHS] Matthew Durham, Mark Hagen, and Alessandro Sisto. Boundaries and automorphisms of hierarchically hyperbolic spaces. *Geom. Topol.*, 21(6):3659–3758, 2017.

[Hae] Thomas Haettel. Hyperbolic rigidity of higher rank lattices. arXiv:1607.02004, 2016.

[Hag] Mark F. Hagen. Weak hyperbolicity of cube complexes and quasi-arboreal groups. *J. Topol.*, 7(2):385–418, 2014.

[HS] Mark F Hagen and Tim Susse. On hierarchical hyperbolicity of cubical groups. arXiv:1609.01313, 2016.

[Hu] Jingyin Huang. Top-dimensional quasiflats in CAT(0) cube complexes. *Geom. Topol.*, 21(4):2281–2352, 2017.

[KK] Sang-Hyun Kim and Thomas Koberda. The geometry of the curve graph of a right-angled Artin group. *Internat. J. Algebra Comput.*, 24(2):121–169, 2014.

[MM1] Howard A. Masur and Yair N. Minsky. Geometry of the complex of curves. I. Hyperbolicity. *Invent. Math.*, 138(1):103–149, 1999.

[MM2] H. A. Masur and Y. N. Minsky. Geometry of the complex of curves. II. Hierarchical structure. *Geom. Funct. Anal.*, 10(4):902–974, 2000.

[MO] Ashot Minasyan and Denis Osin. Acylindrical hyperbolicity of groups acting on trees. *Math. Ann.*, 362(3-4):1055–1105, 2015.

[M1] Sarah C. Mousley. Non-existence of boundary maps for some hierarchically hyperbolic spaces. *To appear in Algebr. Geom. Topol.*, arXiv:1610.07691, 2016.

[M2] Sarah C. Mousley. Exotic limit sets of Teichmüller geodesics in the HHS boundary. arXiv:1704.08645, 2017.

[ST] Alessandro Sisto and Samuel J. Taylor. Largest projections for random walks. *To appear in Math. Res. Lett.*, arXiv:1611.07545, 2016.

[S1] Davide Spriano. Hyperbolic HHS I: Factor Systems and Quasi-convex subgroups. arXiv:1711.10931, 2017.

[S2] Davide Spriano. Hyperbolic HHS II: Graphs of hierarchically hyperbolic groups. arXiv:1801.01850, 2018.

PART THREE

RESEARCH ARTICLES

6

A counterexample to questions about boundaries, stability, and commensurability

Jason Behrstock
CUNY Graduate Center
365 5th Ave.
New York, NY 10016
United States

Abstract

We construct a family of right-angled Coxeter groups that provide counter-examples to questions about the stable boundary of a group, one-endedness of stable subgroups, and the commensurability types of right-angled Coxeter groups.

Introduction

In this note we construct right-angled Coxeter groups with some interesting properties. These examples show that a number of questions in geometric group theory have more nuanced answers than originally expected. In particular, these examples resolve the following questions in the negative:

- (Charney and Sisto): As is the case for right-angled Artin groups, do all (non-relatively hyperbolic) right-angled Coxeter groups have totally disconnected contracting boundary?
- (Taylor): Given that all known quasigeodesically stable subgroups of the mapping class group are virtually free, does it hold in any (non-relatively hyperbolic) group that all quasigeodesically stable subgroups have more than one end?
- (Folk question): If a right-angled Coxeter group has quadratic divergence, must it be virtually a right-angled Artin group?

We describe a family of graphs, any one of which is the presentation graph of a right-angled Coxeter group that provides a counterexample to all three of the above questions. We expect that in special cases, and perhaps in general with appropriate modifications, there are interesting positive answers to these questions; we hope this note will encourage the careful reader to formulate and prove such results.

The construction we give was inspired by thinking about the simplicial boundary for the Croke–Kleiner group, see [BH, Example 5.12] and [Tat2]. In the process we give a quick introduction to a few topics of recent interest in geometric group theory. For further details on these topics see also [ABD, BDM, Ch, CS, Co2, DuT, T].

Acknowledgements

Thanks to Ruth Charney, Alessandro Sisto, and Sam Taylor for sharing with me their interesting questions and to the referee for some helpful comments which led to Remark 6.2.3. Also, thanks to Mark Hagen for many fun discussions relating to topics in this note.

6.1 The construction

Let Γ_n be a graph with $2n$ vertices built in the following inductive way. Start with a pair of vertices a_1, b_1 with no edge between them. Given the graph Γ_{n-1}, obtain the graph Γ_n by adding a new pair of vertices a_n, b_n to the graph Γ_{n-1} and adding four new edges, one connecting each of $\{a_{n-1}, b_{n-1}\}$ to each of $\{a_n, b_n\}$. See Figure 6.1.

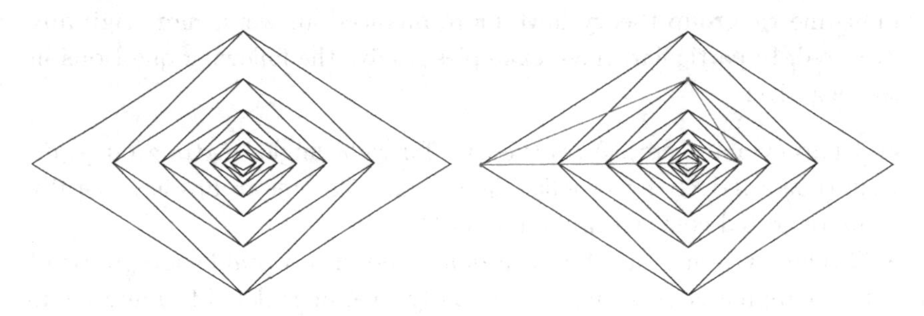

Figure 6.1 The graphs Γ_{14} (left) and Γ (right).

Note that Γ_n is a join if and only if $n \leq 3$. More generally, a_i, a_j are contained in a common join if and only if $|i - j| \leq 2$.

For any $m \geq 5$, choose n sufficiently large so that there exists a set of points $\mathcal{P} = \{p_1, \ldots, p_m\} \subset \Gamma_n$ with the property that for each $1 \leq i < j \leq m$ the points p_i and p_j are not contained in a common join in Γ_n. For example, in Γ_{14} we could choose the vertices $\mathcal{P} = \{a_1, a_4, a_7, a_{10}, a_{13}\}$. For each $1 \leq i < m$ add an edge between p_i and p_{i+1}; also, add an edge between p_m and p_1. Call this new graph Γ. There are many choices of Γ depending on our choices of n, m, and $\mathcal{P}$; for the following any choice will work.

Associated to any graph, one can construct the *right-angled Coxeter group* with that presentation graph. This is the group whose defining presentation is given by: an order-two generator, for each vertex of the graph, and a commutation relation between each of the generators associated to a pair of vertices connected by an edge.

Let W denote the right-angled Coxeter group whose presentation graph is the Γ constructed above. In the next section, we record some key properties about the group W and then, in the final section, apply this to the questions in the Introduction.

6.2 Properties

6.2.1 Quadratic divergence

Proposition 6.2.1 *The group W has quadratic divergence. In particular, this group is not relatively hyperbolic.*

Proof It is easily seen that the graph Γ_n has the property that each vertex is contained in at least one induced square. It is also easy to verify that given any pair of induced squares S, S' in Γ_n, there exists a sequence of induced squares $S = S_1, S_2, S_3, \ldots, S_k = S'$ where for each $1 \leq i < k$ the squares S_i and S_{i+1} share 3 vertices in common. This property, that there are enough squares to chain together any pair of points, is called $\mathcal{CFS}$; it was introduced in [DaT] and studied further in [BF-RHS, L].

Since Γ has the same vertex set as Γ_n, and every induced square in Γ_n is still induced in Γ, it follows that Γ also has the $\mathcal{CFS}$ property.

Given any graph with the $\mathcal{CFS}$ property that is not a join, the associated right-angled Coxeter group has exactly quadratic divergence, see [DaT, Theorem 1.1] and [BF-RHS, Proposition 3.1].

The second statement in the proposition follows from the fact that any relatively hyperbolic group has divergence that is at least exponential [S, Theorem 1.3]. □

6.2.2 Stable surface subgroups

An undistorted subgroup is said to be *(quasi-geodesically) stable* if each pair of points in that subgroup are connected by uniformly Morse quasi-geodesics, see [DuT].

Proposition 6.2.2 *W contains a closed hyperbolic surface subgroup that is stable.*

Proof Recall that in any right-angled Coxeter group an induced subgraph yields a subgroup isomorphic to the right-angled Coxeter group of the associated subgraph. Also note that the right-angled Coxeter group associated to a cycle of length at least 5 is a 2–dimensional hyperbolic orbifold group. Thus the subgraph spanned by $\mathcal{P}$, which is a cycle of length $m \geq 5$, yields a subgroup, H, which is isomorphic to a 2–dimensional hyperbolic orbifold group.

By construction, the subgraph $\mathcal{P}$ doesn't contain any pair of non-adjacent vertices in a common join of Γ (by assumption we use the term join to mean non-degenerate join, in the sense that both parts of the join have diameter at least 2); thus no special subgroup of H is a direct factor in a non-hyperbolic special subgroup of W.

It is proven in [ABD, Theorem B] that a subgroup of a hierarchically hyperbolic group is stable if and only if it is undistorted and has uniformly bounded projections to the curve graph of every proper domain in some hierarchically hyperbolic structure; in turn, this property is equivalent to the orbit of the subgroup being quasi-isometrically embedded in the curve graph of the nest-maximal domain.

A hierarchically hyperbolic structure on right-angled Coxeter groups was built in [BHS]; in that structure the curve graphs are contact graphs, which were first defined in [H], where it was proven that they are all quasi-trees. In the present context, a domain can be thought of as a certain convex subcomplex of the Davis complex arising from special subgroups, and orthogonality holds when the Davis complex contains the direct product of a given such pair.

Since a closed surface group cannot be quasi-isometrically embedded in a quasi-tree, to verify the proposition a different hierarchically hyperbolic structure must be used; for this we now rely on a method developed in [ABD] for modifying structures. The construction given in [ABD, Theorem 3.11] modifies the structure by removing certain domains and augmenting the curve graphs associated to domains in which they are nested. Indeed, passing to the new structure and back, it follows that to verify a given subgroup has uniformly bounded projections in a particular structure, it suffices to verify that the subgroup has uniformly bounded projections to the set of domains $\mathcal{W}$ with the property that for each $W \in \mathcal{W}$, there exists a domain U that is orthogonal to W and has infinite diameter curve graph (since all other domains will be removed in the new structure while only changing the curve graph of the nest-maximal domain).

In the standard hierarchically hyperbolic structure the only domains to which H has unbounded projections are contained in the subcomplex of the Davis complex associated to the special subgroup H. As noted above, since no pair of vertices in $\mathcal{P}$ are contained in a common join in Γ, these domains all have the property that any domains orthogonal to one of them must have uniformly bounded diameter. Thus, applying the results of [ABD] just discussed, it follows that H is stable.

Now take a cover of the orbifold to get the desired closed hyperbolic surface. □

Remark 6.2.3 We note that for the subgroup H constructed in Proposition 6.2.2, it follows easily from [KL, Proposition 3.3] (or, more explicitly, from [T, Theorem 1.1]) that each infinite order element in H acts as a rank-one isometry of the Davis complex of W and is thus Morse. To see that such an argument alone is not enough to prove stability of the subgroup, we note that the authors of [OOS] construct lacunary hyperbolic groups which provide an obstruction. In particular, in [OOS, Theorem 1.12] it is proven that there exist infinite finitely generated non-hyperbolic groups in which every proper subgroup is infinite cyclic and generate uniformly Morse quasi-geodesics.

The next result follows from Proposition 6.2.2 and [Co1, Proposition 4.2].

Corollary 6.2.4 *The right-angled Coxeter group W contains a topologically embedded circle in its Morse boundary.*

6.3 Applications

6.3.1 Morse boundaries

Charney and Sultan introduced a boundary for CAT(0) groups which captures aspects of the negative curvature of the group [CS]. Their construction was then generalized by Cordes to a framework that exists for all finitely generated groups [Co1]; in this general context it is known as the *Morse boundary.* Charney and Sultan built examples of relatively hyperbolic right-angled Coxeter groups whose boundaries are not totally disconnected [CS]. More generally, it is now known that for hyperbolic groups the Morse boundary coincides with the hyperbolic boundary [Co1, Main Theorem (3)]; using this it is easy to produce many examples of hyperbolic and relatively hyperbolic right-angled Coxeter groups whose boundary are not totally disconnected.

On the other hand, the Morse boundary of any right-angled Artin group A_Γ is totally disconnected. The two-dimensional case of this is implicit in [CS]. In general, this fact follows from the fact that the contact graph has totally disconnected boundary (since it is a quasi-tree) and from [CH, Theorem F]. To see this recall that [CH, Theorem F] provides a continuous map from the Morse boundary of A_Γ to the boundary of the contact graph of the universal cover of the Salvetti complex of A_Γ; at which point the result follows since Morse geodesic rays lying at infinite Hausdorff distance cannot fellow-travel in the contact graph, so this continuous boundary map is injective.

Accordingly, Ruth Charney and Alessandro Sisto raised the question of whether outside of the relatively hyperbolic setting, right-angled Coxeter groups all have totally disconnected Morse boundary. The group W constructed above shows the answer is no, since it is not relatively hyperbolic by Proposition 6.2.1 and its boundary is not totally disconnected, by Corollary 6.2.4.

6.3.2 Stable subgroups

Examples of stable subgroups are known both in the mapping class group [B, DuT] and in right-angled Artin groups [KMT]. In both of these classes, all known examples of stable subgroups are virtually free; in the relatively hyperbolic setting on the other hand, it is easy to construct one-ended stable subgroups. Sam Taylor asked whether there exist non-relatively hyperbolic groups with one-ended stable subgroups. The example W is a non-relatively hyperbolic group with one-ended stable subgroups, by Proposition 6.2.1 and Proposition 6.2.2.

6.3.3 Commensurability

A well-known construction of Davis–Januszkiewicz [DJ] shows that every right-angled Artin group is commensurable to some right-angled Coxeter group. The following is a well-known problem:

Question 6.3.1 Which right-angled Coxeter groups are commensurable to right-angled Artin groups?

It is known that any right-angled Artin groups has divergence that is either linear (if it is a direct product) or quadratic, see [BC] or [ABDDY]. Since divergence is invariant under quasi-isometry, and hence under commensurability as well, this puts a constraint on the answer to Question 6.3.1. Several people have raised the question of whether every right-angled Coxeter group with quadratic divergence is quasi-isometric to some right-angled Artin group. The group W shows that the answer is no, since [Co1, Main Theorem (2)] proves that the Morse boundary is invariant under quasi-isometries, but Propositions 6.2.1 and 6.2.2 show that the group W is a right-angled Coxeter group with quadratic divergence whose Morse boundary contains an embedded circle, while the Morse boundary of any right-angled Artin group is totally disconnected.

6.4 Further questions

In [BF-RHS] it was established that for a large range of density functions that asymptotically almost surely the random graph yields a right-angled Coxeter group with quadratic divergence. More generally, it is known that any random graph, with density greater than $1/n$ and bounded away from 1, will asymptotically almost surely contain a large induced polygon. Accordingly, we expect that the proof of Corollary 6.2.4 could be used to verify:

Conjecture 6.4.1 *For any density greater than* $1/n$ *and bounded away from 1, asymptotically almost surely the random right-angled Coxeter group contains circles in its Morse boundary. In particular, it is not virtually a right-angled Artin group.*

The prevalence of right-angled Coxeter groups with quadratic divergence makes the following an appealing (but likely very difficult) question:

Question 6.4.2 Classify right-angled Coxeter groups with quadratic divergence up to quasi-isometry. Classify them up to commensurability.

In light of Proposition 6.2.2 and Remark 6.2.3 it would be interesting if the following was true in general or under some moderate hypotheses:

Question 6.4.3 Let G be a hierarchically hyperbolic group with a subgroup H. If all infinite cyclic subgroups of H are uniformly Morse, does that imply that H is stable?

Note that the lacunary hyperbolic groups discussed in Remark 6.2.3 are not an obstruction to an affirmative answer to Question 6.4.3, since non-hyperbolic lacunary hyperbolic groups are infinitely presented [OOS] while hierarchically hyperbolic groups are all finitely presented [BHS2, Corollary 7.5].

References

[ABD] Carolyn Abbott, Jason Behrstock, and Matthew G Durham. Largest acylindrical actions and stability in hierarchically hyperbolic groups. ARXIV:1705.06219.

[ABDDY] A. Abrams, N. Brady, P. Dani, M. Duchin, and R. Young. Pushing fillings in right-angled Artin groups. *J. Lond. Math. Soc. (2)* **87** (2013), 663–688.

[B] J. Behrstock. Asymptotic geometry of the mapping class group and Teichmüller space. *Geometry & Topology* **10** (2006), 1523–1578.

[BC] J. Behrstock and R. Charney. Divergence and quasimorphisms of right-angled Artin groups. *Math. Ann.* **352** (2012), 339–356.

[BDM] J. Behrstock, C. Druţu, and L. Mosher. Thick metric spaces, relative hyperbolicity, and quasi-isometric rigidity. *Math. Ann.* **344** (2009), 543–595.

[BF-RHS] J. Behrstock, V. Falgas-Ravry, M. Hagen, and T. Susse. Global structural properties of random graphs. *Int. Math. Res. Not.* (2018), no. 5, 1411–1441.

[BH] J. Behrstock and M.F. Hagen. Cubulated groups: thickness, relative hyperbolicity, and simplicial boundaries. *Geometry, Groups, and Dynamics* **10** (2016), 649–707.

[BHS] J. Behrstock, M.F. Hagen, and A. Sisto. Hierarchically hyperbolic spaces I: curve complexes for cubical groups. *Geometry & Topology*, 21:1731–1804, 2017.

[BHS2] J. Behrstock, M.F. Hagen, and A. Sisto. Hierarchically hyperbolic spaces II: combination theorems and the distance formula. To appear in *Pac. J. Math.*

[Ch] Ruth Charney. An introduction to right-angled Artin groups. *Geom. Dedicata* **125** (2007), 141–158.

[CS] Ruth Charney and Harold Sultan. Contracting boundaries of CAT(0) spaces. *J. Topol.* **8** (2015), 93–117.

[Co1] Matthew Cordes. Morse boundaries of proper geodesic metric spaces. ARXIV:1502.04376.

[Co2] Matthew Cordes. A survey on Morse boundaries & stability. ARXIV:1704.07598.

[CH] Matthew Cordes and David Hume. Stability and the Morse boundary. *To appear in Groups, Geometry, and Dynamics* (2016).

[DaT] Pallavi Dani and Anne Thomas. Divergence in right-angled Coxeter groups. *Trans. Amer. Math. Soc.* **367** (2015), 3549–3577.

[DJ] Michael W. Davis and Tadeusz Januszkiewicz. Right-angled Artin groups are commensurable with right-angled Coxeter groups. *J. Pure Appl. Algebra* **153** (2000), 229–235.

[DuT] Matthew Gentry Durham and Samuel J. Taylor. Convex cocompactness and stability in mapping class groups. *Algebr. Geom. Topol.* **15** (2015), 2839–2859.

[H] Mark F. Hagen. Weak hyperbolicity of cube complexes and quasi-arboreal groups. *J. Topol.* **7** (2014), 385–418.

[KL] M. Kapovich and B. Leeb. 3-manifold groups and nonpositive curvature. *Geom. Funct. Anal.* **8** (1998), 841–852.

[KMT] Thomas Koberda, Johanna Mangahas, and Samuel J Taylor. The geometry of purely loxodromic subgroups of right-angled Artin groups. ARXIV:1412.3663.

[L] Ivan Levcovitz. Divergence of CAT(0) Cube Complexes and Coxeter Groups. *Algebr. Geom. Topol.* **18** (2018), 1633–1673.

[OOS] Alexander Yu. Ol′shanskii, Denis V. Osin, and Mark V. Sapir. Lacunary hyperbolic groups. *Geom. Topol.* **13**(2009), 2051–2140. With an appendix by Michael Kapovich and Bruce Kleiner.

[S] A. Sisto. On metric relative hyperbolicity. ARXIV:1210.8081.

[Tat2] Mark F. Hagen. Tattoo. In preparation.

[T] Hung Cong Tran. Purely loxodromic subgroups in right-angled Coxeter groups. ARXIV:1703.09032.

7

A note on the acylindrical hyperbolicity of groups acting on CAT(0) cube complexes

Indira Chatterji
Laboratoire de Mathématiques J.A. Dieudonné
Université de Nice - Sophia Antipolis
06108 Nice Cedex 02
France

Alexandre Martin
Department of Mathematics
Heriot-Watt University
EH14 4AS Edinburgh
United Kingdom

Abstract

We study the acylindrical hyperbolicity of groups acting by isometries on CAT(0) cube complexes, and obtain simple criteria formulated in terms of stabilisers for the action. Namely, we show that a group acting essentially and non-elementarily on a finite dimensional irreducible CAT(0) cube complex is acylindrically hyperbolic if there exist two hyperplanes whose stabilisers intersect along a finite subgroup. We also give further conditions on the geometry of the complex so that the result holds if we only require the existence of a single pair of points whose stabilisers intersect along a finite subgroup.

7.1 Introduction

A group is called *acylindrically hyperbolic* if it is not virtually cyclic and admits an acylindrical action with unbounded orbits on a hyperbolic geodesic metric space. Acylindrically hyperbolic groups form a large class of groups, introduced by Osin [O2], displaying strong hyperbolic-like features: they encompass mapping class groups of hyperbolic surfaces [B], relatively hyperbolic groups [O1], the plane Cremona group [CL, L], and many more. The notion of acylindrical hyperbolicity has gathered a lot of interest in recent years due to the strong algebraic and analytical consequences it implies for the group (we refer to [O2] and details therein).

In this article, we study the acylindrical hyperbolicity of groups acting by isometries on CAT(0) cube complexes. Our goal is to obtain simple acylindrical hyperbolicity criteria for such groups. Recall that an action on a CAT(0) cube complex X is called *essential* if no orbit remains at bounded distance from a half-space of X, and that it is called *non-elementary* if it does not admit a finite orbit in $X \cup \partial_\infty X$.

Our first criterion is formulated in terms of hyperplanes stabilisers, and generalises to finite dimensional CAT(0) cube complexes a criterion due to Minasyan–Osin for actions on simplicial trees [MO]:

Theorem 7.1.1 *Let G be a group acting on an irreducible finite-dimensional CAT(0) cube complex essentially and non-elementarily. If there exist two hyperplanes whose stabilisers intersect along a finite subgroup, then G is acylindrically hyperbolic.*

In the previous theorem, we do not require the two hyperplanes to be disjoint, or even distinct. In particular, the conclusion holds if there exists a hyperplane of X whose stabiliser is *weakly malnormal*, i.e. it intersects some conjugate along a finite subgroup.

It should be noted that the above theorem does not reduce to the aforementioned criterion of Minasyan–Osin, as groups acting on CAT(0) cube complexes do not virtually act on a simplicial tree *a priori*.

Anthony Genevois has also obtained criteria for acylindrical hyperbolicity of a similar flavour, using different tools. In particular, his approach can be used to recover Theorem 7.1.1, see [G2, Remark 21].

We give an application of Theorem 7.1.1 to Artin groups of FC type, suggested to us by Ruth Charney. Artin groups span a large range of groups, and include for instance free groups, braid groups, and free abelian groups, as well as many more exotic groups. Artin groups and their subgroups are a rich source of examples and counterexamples of interesting phenomena in geometry and group theory.

Theorem 7.1.2 *Non-virtually cyclic Artin groups of FC type whose underlying Coxeter graphs have diameter at least* 3 *are acylindrically hyperbolic.*

Artin groups of FC type and the proof of the above theorem will be discussed in Section 7.5. In the case of Artin groups of FC type whose Coxeter graphs have diameter 1, i.e. Artin groups of finite type, Calvez and Wiest [CW] showed, using different techniques, that the quotient of such groups by their centre is acylindrically hyperbolic.

As a corollary of Theorem 7.1.1, we obtain the following results for actions satisfying a weak notion of acylindricity or properness:

Corollary 7.1.3 *Let G be a group acting on an irreducible finite-dimensional CAT(0) cube complex X essentially and non-elementarily. Assume that the action is non-uniformly weakly acylindrical, that is, there exists a constant $L \geq 0$ such that only finitely many elements of G fix two points of X at distance at least L. Then G is acylindrically hyperbolic.*

Corollary 7.1.4 *A group acting essentially, non-elementarily, and with finite vertex stabilisers on a finite dimensional irreducible CAT(0) cube complex is acylindrically hyperbolic.*

Corollary 7.1.3 was already known if the complex is in addition assumed to be *hyperbolic* (see [M1] for the proof in dimension 2 and [G1] for the general case). Corollary 7.1.4 was already known if in addition the action is assumed to be metrically proper, by work of Caprace–Sageev on the existence of rank one isometries of CAT(0) cube complexes [CS]. Anthony Genevois informed us that he found independently similar results, using different techniques [G2].

In Theorem 7.1.1 and its corollaries, the essentiality assumption could be weakened to the assumption that the *essential core* is irreducible, a condition that is *a priori* non-trivial to check. Notice however that the other assumptions cannot be removed. The non-acylindrically hyperbolic group $\mathbb{Z}$ acts properly and essentially, but *elementarily*, on the real line with its standard simplicial structure. The groups of Burger–Mozes provide examples of cocompact lattices in the product of two trees whose associated actions are essential, non-elementary, and without fixed point at infinity, yet these groups are not acylindrically hyperbolic as they are simple [BM]. Thompson's groups V and T act properly, non-elementarily, and without fixed point at infinity on an *infinite dimensional* irreducible CAT(0) cube complex [F], but are not acylindrically hyperbolic as they are also simple.

We also obtain stronger criteria, that allow us to deduce the acylindrical hyperbolicity of a group from information on the stabilisers of a single pair of points (see Theorem 7.4.2 for the general statement). For this, we impose further conditions on the complex: we say that a CAT(0) cube complex is *cocompact* if its automorphism group acts cocompactly on it, and we say that it has *no free face* if every non-maximal cube is contained in at least two maximal cubes. We prove the following:

Theorem 7.1.5 *Let G be a group with an essential and non-elementary action on an irreducible finite-dimensional cocompact CAT(0) cube complex with no free face. If there exist two points whose stabilisers intersect along a finite subgroup, then G is acylindrically hyperbolic.*

To show the acylindrical hyperbolicity of a group G through its action on a geodesic metric space X, a useful criterion introduced by Bestvina–Bromberg–Fujiwara [BBF, Theorem H] is to find an infinite order element of the group whose orbits are strongly contracting and which satisfies the so-called *WPD condition*. However, this condition, formulated in terms of *coarse* stabilisers of pairs of points, is generally cumbersome to check for actions on non-locally finite spaces. In [M2], the second author introduced a different criterion involving a weakening of this condition formulated purely in terms of stabilisers of pairs of points, making it much more tractable (see Theorem 7.2.3 for the exact formulation). The price to pay is to find group elements satisfying a strengthened notion of contraction of their orbits, called *über-contractions*. The second author showed that such contractions abound for actions on non-locally compact spaces under mild assumptions on the stabilisers of vertices, and used it to show the acylindrical hyperbolicity of the tame automorphism group of an affine quadric threefold, a subgroup of the Cremona group $\mathrm{Bir}(\mathbb{P}^3(\mathbb{C}))$. In this article, we provide a different way to construct über-contractions for groups acting on CAT(0) cube complexes, this time using the very rich combinatorial geometry of their hyperplanes. This construction relies heavily on the existence of hyperplanes with strong separation properties, called *über-separated hyperplanes*, and introduced by Chatterji–Fernós–Iozzi [CFI]. Such hyperplanes were used to prove a superrigidity phenomenon for groups acting non-elementarily on CAT(0) cube complexes.

Acknowledgements

The authors warmly thank Talia Fernós for producing the example in Remark 7.4.4, Michah Sageev for discussions related to Proposition 7.4.3, Carolyn Abbott for noticing Theorem 7.1.2 and Ruth Charney for pointing it out to us, along with the explanations of Section 7.5. We also thank Anthony Genevois for remarks on an early version of this article, Ruth Charney and Rose Morris-Wright for spotting a gap in the proof of Theorem 7.1.2, and the anonymous referee for their interesting comments.

The first author is partially supported by the Institut Universitaire de France (IUF) and the ANR GAMME. The second author is partially supported by the Austrian Science Fund (FWF) grant M1810-N25. This work was completed while the two authors were in residence at MSRI during the Fall 2016 "Geometric Group Theory" research program, NSF Grant DMS-1440140. They thank the organisers for the opportunity to work in such a stimulating environment.

7.2 Über-contractions and acylindrical hyperbolicity

Recall that, given a group G acting on a geodesic metric space X, the action is called *acylindrical* if for every $r \geq 0$ there exist constants $L(r), N(r) \geq 0$ such that for all points x, y of X at distance at least $L(r)$,

$$|\{g \in G | d(x, gx) \leq r,\ d(y, gy) \leq r\}| \leq N(r).$$

A group is *acylindrically hyperbolic* if it is not virtually cyclic and admits an acylindrical action with unbounded orbits on a hyperbolic geodesic metric space. Given a group action on an arbitrary geodesic space, the following criterion was introduced by Bestvina–Bromberg–Fujiwara to prove the acylindrical hyperbolicity of the group:

Theorem 7.2.1 ([BBF, Theorem H]) *Let G be a group acting by isometries on a geodesic metric space X. Let g be a group element of infinite order with quasi-isometrically embedded orbits, and assume that the following holds:*

- *the group element g is strongly contracting, that is, there exists a point x of X such that the closest-point projections on the $\langle g \rangle$-orbit of x of the balls of X that are disjoint from $\langle g \rangle x$ have uniformly bounded diameter,*
- *the group element g satisfies the WPD condition, that is, for every $r \geq 0$ and every point x of X, there exists an integer m such that only finitely many elements h of G satisfy $d(x, hx), d(g^m x, hg^m x) \leq r$.*

Then G is either virtually cyclic or acylindrically hyperbolic. □

The following notion, introduced in [M2], provides an easier way to prove the acylindrical hyperbolicity of a group.

Definition 7.2.2 Let X be a geodesic metric space and let h be an isometry of X with quasi-isometrically embedded orbits. A *system of checkpoints* for h is the data of a finite subset S of X, the *checkpoint*, as well as an *error constant* $L \geq 0$ and a quasi-isometry $f : \Lambda := \bigcup_{i \in \mathbb{Z}} h^i S \to \mathbb{R}$ such that we have the following:

Let x, y be points of X and let x', y' be closest-point projections on Λ of x, y respectively. For every *checkpoint* $S_i := h^i S, i \in \mathbb{Z}$, such that:

- S_i *coarsely separates* x' and y' , i.e. $f(x')$ and $f(y')$ lie in different unbounded connected components of $\mathbb{R} \setminus f(S_i)$,
- S_i is at distance at least L from both x' and y',

then every geodesic between x and y meets S_i.

A hyperbolic isometry h of X is *über-contracting*, or is an *über-contraction*, if it admits such a system of checkpoints.

We will be using the following criterion for acylindrical hyperbolicity:

Theorem 7.2.3 (Theorem 1.2 of [M2]) *Let G be a group acting by isometries on a geodesic metric space X. Let $g \in G$ be an infinite order element such that the following holds:*

(i) the group element g is über-contracting with respect to a system of checkpoints $(g^i S)_{i \in \mathbb{Z}}$,
(ii) there exists a constant m_0 such that for every point $s \in S$ and every $m \geq m_0$, only finitely many elements of G fix s and $g^m s$ pointwise.

Then G is either virtually cyclic or acylindrically hyperbolic. □

Note that condition (*ii*) of the previous theorem is a considerable weakening of the WPD condition for g.

This weaker condition has the advantage of involving only stabilisers of pairs of points, which makes it much easier to use.

7.3 Über-separated hyperplanes and the proof of Theorem 7.1.1

We now recall a few basic facts concerning CAT(0) cube complexes, and more precisely the notions of bridges and über-separated pairs, which are crucial to prove Theorems 7.1.1 and 7.1.5. The missing details and proofs can be found in [CFI].

By a slight abuse of notation, we will identify a CAT(0) cube complex with its vertex set endowed with the graph metric coming from its 1-skeleton. Each hyperplane separates the vertex set into two disjoint components, which we refer to as halfspaces.

The following notion was introduced by Behrstock–Charney in [BC]:

Definition 7.3.1 Two parallel hyperplanes of a CAT(0) cube complex are said to be *strongly separated* if no hyperplane is transverse to both. By the usual abuse of terminology, we say that two halfspaces are strongly separated if the corresponding hyperplanes are.

We will need tools introduced by Caprace–Sageev [CS]. Recall that a family of n pairwise crossing hyperplanes divides a CAT(0) cube complex into 2^n regions called *sectors*.

Lemma 7.3.2 (Caprace–Sageev [CS, Proposition 5.1, Double-Skewering Lemma, Lemma 5.2]) *Let X be a finite dimensional irreducible CAT(0) cube complex and let $G \to \mathrm{Aut}(X)$ be a group acting essentially and without fixed points in $X \cup \partial_\infty X$.*

- *(Strong Separation Lemma) Let h_1 be a halfspace of X. Then there exists a halfspace h_2 such that $h_1 \subset h_2$ and such that h_1 and h_2 are strongly separated.*
- *(Double-Skewering Lemma) Let $h_1 \subset h_2$ be two nested halfspaces. Then there exists an element $g \in G$ that double-skewers h_1 and h_2, that is, such that $h_1 \subset h_2 \subset gh_1$.*
- *(Sector Lemma) Let $\hat{h}_1$, $\hat{h}_2$ be two transverse hyperplanes. Then we can choose two disjoint hyperplanes $\hat{h}_3$ and $\hat{h}_4$ that are contained in opposite sectors determined by $\hat{h}_1$ and $\hat{h}_2$.* □

We will need a finer notion of strong separation of halfspaces, which is less standard but will be key to our work.

Definition 7.3.3 Two strongly separated halfspaces h_1 and h_2 are said to be an *über-separated pair* if for any two halfspaces k_1, k_2 with the property that h_i and k_i are transverse for $i = 1, 2$, then k_1 and k_2 are parallel. We say that two strongly separated hyperplanes are *über-separated* if their halfspaces are.

Note that pairs of über-separated hyperplanes correspond exactly to pairs of hyperplanes at distance at least 4 in the intersection graph.

Remark 7.3.4 If $h \subset k \subset \ell$ are pairwise strongly separated halfspaces, then h and ℓ are über-separated.

Notice that über-separated pairs are in particular strongly separated and hence they do not exist in the reducible case by [CS, Proposition 5.1]. The existence of über-separated hyperplanes is a consequence of the Double-Skewering Lemma 7.3.2. We will need the following lemma, which is a direct consequence of the Double-Skewering Lemma 7.3.2 and [CFI, Lemma 2.14]:

Lemma 7.3.5 *Let X be a finite dimensional irreducible CAT(0) cube complex and $G \to \mathrm{Aut}(X)$ a group acting essentially and non-elementarily. Given any two parallel hyperplanes $\hat{h}_1$ and $\hat{h}_2$, there exists $g \in G$ that double-skewers $\hat{h}_1$ and $\hat{h}_2$ and such that $\hat{h}_1$ and $g\hat{h}_1$ are über-separated.* □

For two points x, y of X, we denote by $\mathcal{I}(x,y)$ the interval between x and y, that is, the union of all the geodesics between x and y. Recall that intervals are finite, as they only depend on the (finite) set of hyperplanes separating the given pair of points.

Definition 7.3.6 Let $h_1 \subset h_2$ be a nested pair of halfspaces. The *(combinatorial) bridge* between $\hat{h}_1$ and $\hat{h}_2$, denoted $b(\hat{h}_1, \hat{h}_2)$, is the union of all the geodesics between points $x_1 \in h_1$ and $x_2 \in h_2^*$ minimizing the distance between h_1 and h_2^*.

Lemma 7.3.7 (Lemma 2.18 and 2.24 [CFI]) *If $h_1 \subset h_2$ are a pair of nested halfspaces, strongly separated, there exists a unique pair of points of $h_1 \times h_2^*$, called the* gates *of the bridge, that minimizes the distance between h_1 and h_2^*, i.e. there exist points $x_1 \in h_1$ and $x_2 \in h_2^*$ such that*

$$b(\hat{h}_1, \hat{h}_2) = \mathcal{I}(x_1, x_2).$$

In particular, the bridge between two strongly separated hyperplanes is finite. Moreover, the following holds for any $y_1 \in h_1$ and $y_2 \in h_2^*$.

$$d(y_1, y_2) = d(y_1, x_1) + d(x_1, x_2) + d(x_2, y_2).$$

□

The following is a very important feature of über-separated pairs.

Lemma 7.3.8 (Proof of Lemma 3.5 of [CFI]) *Let $h_1 \subset h_2$ be an über-separated pair of halfspaces, $x \in h_1$ and $y \in h_2^*$. Then every geodesic between x and y meets the bridge $b(\hat{h}_1, \hat{h}_2)$.* □

The following lemma explains the relationship between über-separated pairs and über-contractions.

Lemma 7.3.9 *Let $h_1 \subset h_2$ be an über-separated pair of halfspaces, and $g \in G$ an element that double-skewers h_1 and h_2. Then g is an über-contraction.*

Proof Let B denote the bridge between h_1 and gh_1. We will show that the collection of the g-translates of B forms a system of checkpoints. Notice that since the bridge is an interval, it is finite even though the complex is not assumed to be locally finite, hence all the checkpoints $S_n := g^n B$ are finite. Set $\Lambda := \bigcup_{n \in \mathbb{Z}} S_n$. Let $Y := h_1^* \cap gh_1$ and $Y_n := g^n Y$ for every $n \in \mathbb{Z}$.

For every point $x \in Y_n$, we have that its closest-point projections on Λ are in $S_{n-1} \cup S_n \cup S_{n+1}$: Indeed, since h_1 and gh_1 are über-separated, it follows from Lemma 7.3.8 that a geodesic between x and a point $x' \in Y_{n'}$ with $n' \geq n+2$ ($n' \leq n-2$ respectively) meets the bridge S_{n+1} (S_{n-1} respectively). Moreover, for every $x \in X$ that projects on Λ to a point of S_n, we have that $x \in Y_{n-1} \cup Y_n \cup Y_{n+1}$ for the same reasons. Thus, for $x, y \in X$ that project on Λ to points $x' \in S_n$ and $y' \in S_m$ respectively with $m \geq n+4$, it follows that x and y are separated by the hyperplanes $g^{n+2}\hat{h}_1, \dots, g^{m-1}\hat{h}_1$, hence every geodesic between x and y meets each checkpoint $S_{n+2}, \dots, S_{m-2}$ by Lemma 7.3.8.

By a result of Haglund [H, Theorem 1.4], g admits a combinatorial axis Λ_g. As Λ and Λ_g stay at bounded distance from one another, it follows easily from the discussion of the previous paragraph that Λ_g is contained in Λ. Since B is finite, the closest-point projection $\Lambda \to \Lambda_g$ yields a quasi-isometry $\Lambda \to \mathbb{R}$, and it is now straightforward to check that g is an über-contraction. □

Proof of Theorem 7.1.1 Let G be a group acting essentially and non-elementarily on an irreducible finite-dimensional CAT(0) cube complex X and assume that there exist two hyperplanes whose stabilisers intersect along a finite subgroup. By Proposition 7.2.3, it is enough to construct an element of G satisfying conditions (i) and (ii) of Proposition 7.2.3.

Let $\hat{h}_1$ and $\hat{h}_2$ be two hyperplanes whose stabilisers intersect along a finite subgroup. We start by showing that we can assume that $\hat{h}_1$ and $\hat{h}_2$ are disjoint. By the Sector Lemma 7.3.2, we choose two disjoint hyperplanes $\hat{h}_3$ and $\hat{h}_4$ that are contained in opposite sectors determined by $\hat{h}_1$ and $\hat{h}_2$. Up to applying the Strong Separation Lemma 7.3.2, we can further assume that $\hat{h}_3$ and $\hat{h}_4$ are strongly separated. Let $H := \mathrm{Stab}(\hat{h}_3) \cap \mathrm{Stab}(\hat{h}_4)$. Then H stabilises the bridge between $\hat{h}_3$ and $\hat{h}_4$, which is a single interval by Lemma 7.3.7 since $\hat{h}_3$ and $\hat{h}_4$ are strongly separated. As intervals are finite, H virtually fixes a geodesic between $\hat{h}_3$

and $\hat{h}_4$. As $\hat{h}_3$ and $\hat{h}_4$ are in opposite sectors determined by $\hat{h}_1$ and $\hat{h}_2$, such a geodesic crosses both $\hat{h}_1$ and $\hat{h}_2$. Thus, H is virtually contained in $\mathrm{Stab}(\hat{h}_1) \cap \mathrm{Stab}(\hat{h}_2)$, hence H is finite, which is what we wanted.

Thus, let $\hat{h}_1$ and $\hat{h}_2$ be two disjoint hyperplanes whose stabilisers intersect along a finite subgroup. By Lemma 7.3.5, we choose an element g of G that double-skewers $\hat{h}_1$ and $\hat{h}_2$, and such that $\hat{h}_1$ and $g\hat{h}_1$ are über-separated. Then according to Lemma 7.3.9 the element g is an über-contraction, proving (i). Let x be a point of the bridge between $\hat{h}_1$ and $\hat{h}_2$, and choose a geodesic γ between x and g^2x. We have that γ crosses $g\hat{h}_1$ and $g\hat{h}_2$. Since intervals in a finite dimensional CAT(0) cube complex are finite, a subgroup fixing both x and g^2x virtually fixes γ pointwise. It follows that $\mathrm{Stab}(x) \cap \mathrm{Stab}(g^2x)$ is virtually contained in a conjugate of $\mathrm{Stab}(\hat{h}_1) \cap \mathrm{Stab}(\hat{h}_2)$, which is finite. This proves (ii), and Proposition 7.2.3 implies that G is either virtually cyclic or acylindrically hyperbolic. The action being non-elementary, the virtually cyclic case is automatically ruled out, which concludes the proof. □

Proof of Corollaries 7.1.3 and 7.1.4 Since Corollary 7.1.4 is a direct consequence of Corollary 7.1.3, let us assume that the group G acts essentially, non-elementarily, and non-uniformly weakly acylindrically (with a constant L as in the statement) on the irreducible finite-dimensional CAT(0) cube complex X. By the Strong Separation Lemma 7.3.2, choose two disjoint hyperplanes $\hat{h}_1$ and $\hat{h}_2$. This implies that the combinatorial bridge between $\hat{h}_1$ and $\hat{h}_2$ is not reduced to a point. Moreover, the subgroup $\mathrm{Stab}(\hat{h}_1) \cap \mathrm{Stab}(\hat{h}_2)$ is virtually contained in the pointwise stabiliser of the finite bridge between $\hat{h}_1$ and $\hat{h}_2$ by Lemma 7.3.7.

By the Double-Skewering Lemma 7.3.2, choose a group element g that skewers both $\hat{h}_1$ and $\hat{h}_2$. For n large enough, $\hat{h}_1$ and $g^n\hat{h}_1$ are über-separated, and the distance between $\hat{h}_1$ and $g^n\hat{h}_1$ becomes greater than L by Lemma 7.3.7. In particular, $\mathrm{Stab}(\hat{h}_1) \cap \mathrm{Stab}(g^n\hat{h}_1)$ virtually fixes a path of length L, hence $\mathrm{Stab}(\hat{h}_1) \cap \mathrm{Stab}(g^n\hat{h}_1)$ is finite by weak acylindricity. Corollary 7.1.3 now follows from Theorem 7.1.1. □

Theorem 7.1.1 allows for a very simple geometric proof of the acylindrical hyperbolicity of certain groups:

Example 7.3.10 The Higman group on $n \geq 4$ generators, defined by the following presentation:

$$H_n := \langle a_i, i \in \mathbb{Z}/n\mathbb{Z} \mid a_i a_{i+1} a_i^{-1} = a_{i+1}^2, i \in \mathbb{Z}/n\mathbb{Z} \rangle,$$

was proved to be acylindrically hyperbolic by Minasyan–Osin [MO], by means of its action on the Bass–Serre tree associated to some splitting. The Higman graph also acts cocompactly, essentially, and non-elementarily on a CAT(0) square complex associated to its standard presentation. That CAT(0) square complex is irreducible, as links of vertices are easily shown not to be complete bipartite graphs (see [M3, Corollary 4.6]). Since square stabilisers are trivial by construction, the stabilisers of two crossing hyperplanes intersect along a finite (actually, trivial) subgroup. Thus Theorem 7.1.1 applies.

7.4 Proof of Theorem 7.1.5

Definition 7.4.1 Let X be a CAT(0) cube complex and C, C' two cubes of X. We say that C and C' *separate a pair of hyperplanes* if there exist two hyperplanes $\hat{h}, \hat{h}'$ of X such that each hyperplane defined by an edge of $C \cup C'$ separates $\hat{h}$ and $\hat{h}'$.

Theorem 7.1.5 will be a consequence of the following more general result:

Theorem 7.4.2 *Let G be a group acting on a finite dimensional irreducible CAT(0) cube complex essentially and non-elementarily. Assume that there exist two maximal cubes C, C' of X whose stabilisers intersect along a finite subgroup, and such that C and C' separate a pair of hyperplanes. Then G is acylindrically hyperbolic.*

Proof Let $\mathcal{H}_{C,C'}$ be the set of hyperplanes defined by the edges of $C \cup C'$. Since C and C' are maximal cubes, we have that $\bigcap_{\hat{k} \in \mathcal{H}_{C,C'}} \mathrm{Stab}(\hat{k})$ is contained in $\mathrm{Stab}(C) \cap \mathrm{Stab}(C')$, and it follows that $\bigcap_{\hat{k} \in \mathcal{H}_{C,C'}} \mathrm{Stab}(\hat{k})$ is finite.

By assumption, choose two disjoint halfspaces h, h' separated by each hyperplane of $\mathcal{H}_{C,C'}$. Up to applying the Strong Separation Lemma 7.3.2, we can further assume that h and h' are strongly separated. In particular, since the bridge between h and h' is finite by Lemma 7.3.7, it follows that $\mathrm{Stab}(\hat{h}) \cap \mathrm{Stab}(\hat{h}')$ is virtually contained in the pointwise stabiliser of a geodesic between the two gates of the bridge $b(\hat{h}, \hat{h}')$. As such a geodesic crosses each hyperplane of $\mathcal{H}_{C,C'}$ by construction of h, h', it follows that $\mathrm{Stab}(\hat{h}) \cap \mathrm{Stab}(\hat{h}')$ is virtually contained in $\bigcap_{\hat{k} \in \mathcal{H}_{C,C'}} \mathrm{Stab}(\hat{k})$, which is finite by the above argument. Thus $\mathrm{Stab}(h) \cap \mathrm{Stab}(h')$ is finite, and we conclude with Theorem 7.1.1. □

The following proposition gives a class of examples of CAT(0) cube complexes where each pair of cubes separates a pair of hyperplanes:

Proposition 7.4.3 *Let X be an irreducible CAT(0) cube complex without free face and such that* $\mathrm{Aut}(X)$ *acts cocompactly on X. Then each pair of cubes separates a pair of hyperplanes.*

Remark 7.4.4 The following example, due to Talia Fernós, shows that the no-free-face assumption is necessary in Proposition 7.4.3. Take a 3-dimensional cube $[0,1]^3$ and glue an edge to each of the vertices $(1,0,0), (0,1,0), (0,0,1)$ and $(1,1,1)$ to get a spiked cube. Then glue infinitely many of these spiked cubes in a tree-like way, to obtain a cocompact CAT(0) cube complex X which is quasi-isometric to a tree. Then none of the 3-cubes of that complex is separating a pair of hyperplanes. Indeed, each 3-cube has three hyperplanes defining it, hence it defines eight sectors, out of which four contain hyperplanes and four are reduced to a single point, but among the sectors containing hyperplanes, no two are opposite.

Before proving Proposition 7.4.3, we start by a simple observation:

Lemma 7.4.5 *A CAT(0) cube complex with no free face is geodesically complete, that is, every finite geodesic can be extended to a bi-infinite geodesic.*

Proof Let γ be a finite geodesic defined by a sequence $e_1, \dots, e_n$ of edges of X and let v be the terminal vertex of that finite geodesic. We will show that we can extend it by one edge. Let E denote the set of edges of X, containing v and such that γ followed by $e \in E$ is not a geodesic. Then every $e \in E$ has to be parallel to one of the edges e_i defining γ, that is, every $e \in E$ defines a hyperplane $\hat{h}_e$ crossed by γ. Moreover, the map $e \mapsto \hat{h}_e$ is injective as two adjacent edges cannot belong to the same hyperplane, and any two $\hat{h}_e$, $\hat{h}_{e'}$ intersect. Indeed, for an edge e to belong to E means that γ has been travelling on the carrier of the hyperplane defined by some e_i after having crossed e_i, and that can be done simultaneously for several hyperplanes only when they cross each other. Hence E defines a cube in X, and since there are no free faces there is an edge e containing v and that does not belong to E, allowing us to extend by one edge the geodesic γ. □

We will also need the following strengthening of the Sector Lemma:

Lemma 7.4.6 (Strong Sector Lemma) *Let X be a CAT(0) cube complex* *with no free face, and assume that the automorphism group of X acts cocompactly on X. Then each sector determined by a finite family of pairwise crossing hyperplanes contains a hyperplane.*

Proof We prove the result by induction on the number $n \geq 2$ of pairwise crossing hyperplanes. For $n = 2$, the result follows from [S, Proposition 3.3], since a CAT(0) cube complex with no free face is geodesically complete by Lemma 7.4.5. Let $\hat{h}_1, \ldots, \hat{h}_n$ be a family of pairwise crossing hyperplanes, and let h_i be a halfspace associated to $\hat{h}_i$ for every i. We want to construct a hyperplane contained in $\bigcap_i h_i$.

By Helly's theorem, the family $(\hat{h}_i \cap \hat{h}_1)_{i \neq 1}$ defines a family of pairwise crossing hyperplanes of $\hat{h}_1$. Note that $\hat{h}_1$ also satisfies the property of having no free face, and the action of $\mathrm{Stab}(\hat{h}_1)$ on $\hat{h}_1$ is cocompact, as the same holds for the action of $\mathrm{Aut}(X)$ on X. Thus, one can apply the induction hypothesis to find a hyperplane $\hat{h}'$ of $\hat{h}_1$ contained in the sector $\bigcap_{i \neq 1}(\hat{h}_1 \cap h_i)$. This defines a hyperplane of X, which we denote $\hat{h}$. Since $\hat{h}'$ is disjoint from the $\hat{h}_1 \cap \hat{h}_i$ for $i \neq 1$, it follows from Helly's theorem that $\hat{h}$ is disjoint from the $\hat{h}_i$ for $i \neq 1$. Let h be the halfspace of $\hat{h}$ contained in $\bigcap_{i \neq 1} h_i$. Since $\hat{h}$ and $\hat{h}_1$ cross, we can choose by the induction hypothesis a hyperplane contained in the sector $h \cap h_1$, hence in $\bigcap_i h_i$. □

Proof of Proposition 7.4.3 It is enough to prove the proposition when C and C' are maximal. Choose x and x' vertices of C, C' respectively that maximize the distance between vertices of C and C'. Let $\mathcal{H}_C$, $\mathcal{H}_{C'}$ be the family of hyperplanes defined by an edge of C, C' respectively. By the Strong Sector Lemma 7.4.6, each sector determined by $\mathcal{H}_C$ or $\mathcal{H}_{C'}$ contains a hyperplane. Thus, choose a hyperplane $\hat{h}$ ($\hat{h}'$ respectively) in the unique sector determined by $\mathcal{H}_C$ ($\mathcal{H}_{C'}$ respectively) that contains x (x' respectively). By construction of x and x', we have that x and x' are in opposite sectors defined by $\mathcal{H}_C$, and in opposite sectors determined by $\mathcal{H}_{C'}$. In particular, every hyperplane of $\mathcal{H}_{C,C'}$ separates x and x', hence separates $\hat{h}$ and $\hat{h}'$. □

Proof of Theorem 7.1.5 Theorem 7.1.5 is a direct consequence of Theorem 7.4.2 and Proposition 7.4.3. □

Theorem 7.1.5 allows for a very simple geometric proof of the acylindrical hyperbolicity of certain groups:

Example 7.4.7 The group of tame automorphisms of an affine quadric threefold, a subgroup of the 3-dimensional Cremona group $\mathrm{Bir}(\mathbb{P}^3(\mathbb{C}))$, acts cocompactly, essentially and non-elementarily on a hyperbolic CAT(0) cube complex without free face [BFL]. The second author showed that there exist two cubes whose stabilisers intersect along a finite subgroup [M2, Proof of Theorem 3.1], hence Theorem 7.1.5 applies.

7.5 Artin groups of type FC

We now give an application of our results to the class of Artin groups of FC type, studied by Charney and Davis in [CD] and which we now describe. Recall that a *Coxeter graph* is a finite, simplicial graph Γ with vertex set S and edges labelled by integers greater than or equal to 2. The label of the edge connecting two vertices s and t is denoted $m(s,t)$, and we set $m(s,t) = \infty$ if s and t are not connected by an edge. The *Artin group associated to a Coxeter graph* Γ is the group given by the presentation:

$$A = \left\langle S \mid \{ \underbrace{sts\dots}_{m(s,t)} = \underbrace{tst\dots}_{m(s,t)} \ : \ s,t \text{ connected by an edge labelled } m(s,t)\} \right\rangle.$$

Adding the extra relations $s^2 = 1$ for all $s \in S$, we obtain a Coxeter group W as a quotient of A. We say that A is *finite type* if the associated Coxeter group W is finite. It was shown in [vdL] that if $T \subseteq S$, the subgroup A_T generated by T is isomorphic to the Artin group associated to the full subgraph of Γ spanned by T. Such subgroups are called *special subgroups* of A. Following Charney–Davis, we say that an Artin group is of *FC type* if every complete subgraph of the Coxeter graph Γ generates a special subgroup of finite type.

Given an Artin group A, the *Deligne complex* $\mathcal{D}_A$ is the cubical complex defined as follows: vertices of $\mathcal{D}_A$ correspond to cosets aA_T, where $a \in A$ and $T \subseteq S$. Note that we allow $T = \varnothing$, in which case $aA_T = \{a\}$. The 1-skeleton $\mathcal{D}_A^1$ of $\mathcal{D}_A$ is obtained by putting an edge between cosets of the form aA_T and $aA_{T'}$ when $T' = T \cup \{t'\}$ for some $t' \in S \setminus T$. In particular, each edge of $\mathcal{D}_A$ is labelled by an element of S. Finally, $\mathcal{D}_A$ is obtained by "filling the cubes", that is, by gluing a k-cube whenever $\mathcal{D}_A^1$ contains a subgraph isomorphic to the 1-skeleton of a k-cube.

In [CD] Theorem 4.3.5 Charney and Davis show that the Deligne complex $\mathcal{D}_A$ of an Artin group A is a CAT(0) cube complex if and only if A is of FC type. Edges and hyperplanes in this CAT(0) cube complex are labelled by elements of S. Moreover, each hyperplane is a translate of a hyperplane $\hat{h}_s$ for some $s \in S$, where $\hat{h}_s$ denotes the hyperplane defined by the edge between $\{1\}$ and $A_{\{s\}}$. It is straightforward to check that the stabiliser of $\hat{h}_s$ is the special subgroup $A_{lk(s)}$ of A determined by the link of the vertex s in Γ.

The *dimension* of an Artin group is the maximum cardinality of a subset $T \subseteq S$ such that A_T is finite type, and it is equal to the dimension of $\mathcal{D}_A$. In particular, $\mathcal{D}_A$ is finite dimensional since the graph Γ is finite. The Artin group A acts by left multiplication on the aforementioned cosets, and hence acts by isometries on $\mathcal{D}_A$. Since Γ is finite, the action is cocompact. The stabilizer of a vertex aA_T of $\mathcal{D}_A$ is the subgroup aA_Ta^{-1}. In particular, the action is not proper if Γ is non-empty.

The action satisfies the following:

Proposition 7.5.1 *Let A be an Artin group of type FC associated to a Coxeter graph Γ of diameter at least 3. Then the Deligne complex $\mathcal{D}_A$ is irreducible and the action of A on $\mathcal{D}_A$ is essential and non-elementary.*

The proof will take up most of this section. We can assume that Γ is connected, for otherwise A is a free product, $\mathcal{D}_A$ has a structure of a tree of spaces, and the result follows. We check separately the irreducibility of the Deligne complex, the essentiality of the action, and the non-elementarity of the action.

Lemma 7.5.2 *The CAT(0) cube complex $\mathcal{D}_A$ is irreducible.*

Proof Consider the vertex $\{1\}$ of $\mathcal{D}_A$, where 1 denotes the identity element. The labelling of the edges of $\mathcal{D}_A$ yields a surjective map $lk(\{1\}) \to \Gamma$. As Γ has diameter at least 3, so does $lk(\{1\})$, and it follows that $lk(\{1\})$ is not a join, hence $\mathcal{D}_A$ is an irreducible CAT(0) cube complex. □

Lemma 7.5.3 *The action of A on $\mathcal{D}_A$ is essential.*

Proof Since the action is cocompact, it is enough to show that each hyperplane is essential, that is, no half-space is contained in a neighbourhood of the other halfspace. As each hyperplane is a translate of some $\hat{h}_s$, it is enough to show that this is the case for hyperplanes of the form $\hat{h}_s$, where $s \in S$. Let s_0 be a vertex of Γ. Since Γ is connected and has diameter at least 3, we can find two distinct vertices s_1, s_2 of Γ such that s_0, s_1, s_2 defines a geodesic of Γ. Then the hyperplane $\hat{h}_{s_1}$ is in particular stabilized by $A_{\{s_0,s_2\}}$. Thus, to show that $\hat{h}_{s_0}$ is essential, it is enough to show that $\hat{h}_{s_1}$ is unbounded and crosses $\hat{h}_{s_0}$. Let C_{s_0,s_1}, C_{s_1,s_2} be the squares of $\mathcal{D}_A$ containing the vertices $\{1\}, A_{\{s_0\}}, A_{\{s_1\}}$ and $\{1\}, A_{\{s_1\}}, A_{\{s_2\}}$ respectively. Notice that the edge between $A_{\{s_0\}}$ and $A_{\{s_0,s_1\}}$ has stabiliser

$A_{\{s_0\}}$, and the edge between $A_{\{s_2\}}$ and $A_{\{s_1,s_2\}}$ has stabiliser $A_{\{s_2\}}$. Thus, the $A_{\{s_0,s_2\}}$-orbit Y of $C_{s_0,s_1} \cup C_{s_1,s_2}$ defines a subcomplex of $\mathcal{D}_A$ that is convex for the CAT(0) metric and quasi-isometric to a tree. Moreover, every point of Y contained in a half-space of $\hat{h}_{s_0}$ projects on the other half-space to a point of $C_{s_0,s_1} \cup C_{s_1,s_2}$. In particular, each half-space of $\hat{h}_{s_0}$ contains points of Y arbitrarily far away from the other half-space, hence $\hat{h}_{s_0}$ is essential. □

Lemma 7.5.4 *The action of A on $\mathcal{D}_A$ is non-elementary.*

The proof of this lemma requires some preliminary work. Since Γ has diameter at least 3, let s_0, s_1, s_2, s_3 be a geodesic of Γ.

Lemma 7.5.5 *Let $g := s_0 s_3$. Then g is a hyperbolic element and admits an axis Λ_g (for the CAT(0) metric) that is a reunion of geodesic segments such that two consecutive segments make an angle strictly greater than π (for the angular distance on the link).*

Proof For $i = 0$ or 3, we denote by e_i the edge between the vertices $\{1\}$ and $A_{\{s_i\}}$ of $\mathcal{D}_A$. Let $Y := s_0^{-1}e_0 \cup e_0 \cup e_3 \cup s_3 e_3$. Then $\Lambda_g := \bigcup_{n \in \mathbb{Z}} g^n Y$ is a CAT(0) geodesic, and an axis for g. Indeed, we have the following properties of angles between consecutive edges:

- The angle (for the angular distance on the link) between e_0 and e_3 is $3\pi/2 > \pi$ by construction of s_0, s_1, s_2, s_3.
- The angle between e_0 and $s_0^{-1}e_0$ (between e_3 and $s_3 e_3$ respectively) is π: indeed, the angle is at most π by construction, and if the angle were $\pi/2$, then since $\mathcal{D}_A$ is a CAT(0) cube complex, e_i and $s_i e_i$ would be two adjacent edges of a 3-cube with label e_i, which is impossible by construction of $\mathcal{D}_A$.

Thus Λ_g is a local geodesic, hence a global geodesic. Moreover, Λ_g is clearly invariant under the action of $\langle g \rangle$, and the angle made by Λ_g at every $\langle g \rangle$-translate of the vertex $\{1\} \in \mathcal{D}_A$ is $3\pi/2$. □

To show that g is a rank-one element, we need the following modified version of the Flat Strip Theorem:

Lemma 7.5.6 (Flat "Half-strip" Theorem) *Let Y be a CAT(0) space, let Λ be a geodesic line, and let h be an isometry of Y preserving Λ. Let $y \in \partial Y \setminus \partial \Lambda$ and let γ be a geodesic ray from a point x of Λ to y that meets Λ in exactly one point. If γ and $h\gamma$ are asymptotic, then the convex hull of $\gamma \cup h\gamma$ isometrically embeds in $\mathbb{R}^2$ with its standard CAT(0) metric.*

Proof Let us construct a 'double' of Y as follows. Let $o \in \Lambda$ be the midpoint between x and hx. Such a choice allows to define (uniquely) a reflection ψ of Λ across o which is an isometry of Λ. We then define

$$Y' = (Y \sqcup Y)/\psi,$$

i.e. the space obtained from two copies of Y by identifying the two copies of Λ using the reflection ψ. As a convention, we denote these copies by Y_1 and Y_2, we use the subscript $\cdots_i$ to indicate to which copy of Y the object belongs, and we identify Y with the subspace Y_1 of Y'. The space Y' is again a CAT(0) space (as it is obtained from two CAT(0) spaces by identifying two convex subspaces along an isometry). Note that $\gamma \subset Y$ is a sub-ray of $\gamma_1 \cup (h\gamma)_2$ and $h\gamma \subset Y$ is a sub-ray of $(h\gamma)_1 \cup \gamma_2$. Moreover, we have constructed Y' so that $\gamma_1 \cup (h\gamma)_2$ and $(h\gamma)_1 \cup \gamma_2$ are local geodesics of Y', hence global geodesics of Y'. As these geodesic lines are asymptotic by construction, it follows that their geodesic hull is a flat strip by the Flat Strip Theorem [BH, Theorem II.2.13]. In particular, the geodesic hull of $\gamma \cup h\gamma$ in Y isometrically embeds in $\mathbb{R}^2$. □

We can now prove:

Lemma 7.5.7 *The only points of $\partial \mathcal{D}_A$ fixed by g are its two limit points $g^{+\infty}, g^{-\infty} \in \partial \Lambda_g$.*

Proof Let $z \in \partial \mathcal{D}_A \setminus \{g^{+\infty}, g^{-\infty}\}$ and let γ' be a geodesic ray from Λ_g to z that meets Λ_g in exactly one point. If $gz = z$, then γ' and $g^2\gamma'$ are asymptotic geodesic rays, hence the convex hull H of $\gamma' \cup g^2\gamma'$ isometrically embeds in $\mathbb{R}^2$ by the Flat Half-strip Theorem 7.5.6. But γ and $g^2\gamma'$ both meet Λ_g in exactly one point, and by construction H contains two geodesic subsegments of Λ_g which make an angle $3\pi/2$ for the angular distance on the link. This yields the desired contradiction. □

Proof of Lemma 7.5.4 First notice that one could have proved the above lemma for the element $h := s_0^2 s_3$ by applying exactly the same reasoning. If A were to admit a finite orbit at infinity, then some power of g and some power of h would fix a common point at infinity. But the axes of g and h cannot be asymptotic, for otherwise, being already distinct, Λ_g and Λ_h would bound a flat strip of positive width by the Flat Strip Theorem 7.5.6, and the same reasoning as above would yield a contradiction. □

Proof of Proposition 7.5.1 This is a direct consequence of Lemmas 7.5.2, 7.5.3, and 7.5.4. □

We are now ready to apply Theorem 7.1.1 to the action of A on $\mathcal{D}_A$ in order to complete the proof of Theorem 7.1.2.

Proof of Theorem 7.1.2 It is enough to consider the case where Γ is connected, for otherwise the group is a free product. According to the previous lemma, the action of A on its Deligne complex $\mathcal{D}_A$ is essential and non-elementary. In order to apply Theorem 7.1.1 to the action of A on $\mathcal{D}_A$, we choose two vertices s, t of S with disjoint links, which is possible since Γ has diameter at least 3. It follows from the aforementioned result of [vdL] that $A_{lk(s)} \cap A_{lk(t)} = A_{\varnothing} = \{1\}$. Thus, the hyperplanes $\hat{h}_s$ and $\hat{h}_t$ have stabilisers that intersect trivially, hence Theorem 7.1.1 applies and A is acylindrically hyperbolic. □

References

[BC] J. Behrstock and R. Charney. Divergence and quasimorphisms of right-angled Artin groups. *Math. Ann.*, 352(2):339–356, 2012.

[BBF] M. Bestvina, K. Bromberg, and K. Fujiwara. Constructing group actions on quasi-trees and applications to mapping class groups. *Publ. Math. Inst. Hautes Études Sci.*, 122:1–64, 2015.

[BFL] C. Bisi, J.-P. Furter, and S. Lamy. The tame automorphism group of an affine quadric threefold acting on a square complex. *Journal de l'Ecole Polytechnique*, 1:161–223, 2014.

[B] B.H. Bowditch. Tight geodesics in the curve complex. *Invent. Math.*, 171(2):281–300, 2008.

[BH] M. R. Bridson and A. Haefliger. *Metric spaces of non-positive curvature*, volume 319 of *Grundlehren der Mathematischen Wissenschaften [Fundamental Principles of Mathematical Sciences]*. Springer-Verlag, Berlin, 1999.

[BM] M. Burger and S. Mozes. Lattices in product of trees. *Inst. Hautes Études Sci. Publ. Math.*, (92):151–194 (2001), 2000.

[CW] M. Calvez and B. Wiest. Acylindrical hyperbolicity and Artin-Tits groups of spherical type. *Geom. Dedicata.*, 191:199–215, 2017.

[CL] S. Cantat and S. Lamy. Normal subgroups in the Cremona group. *Acta Math.*, 210(1):31–94, 2013. With an appendix by Yves de Cornulier.

[CS] P.-E. Caprace and M. Sageev. Rank rigidity for CAT(0) cube complexes. *Geom. Funct. Anal.*, 21(4):851–891, 2011.

[CD] R. Charney and M. W. Davis. The $K(\pi, 1)$-problem for hyperplane complements associated to infinite reflection groups. *J. Amer. Math. Soc.*, 8(3):597–627, 1995.

[CFI] I. Chatterji, T. Fernós, and A. Iozzi. The median class and superrigidity of actions on CAT(0) cube complexes. *J. Topol.*, 9(2):349–400, 2016.

[F] D. Farley. The action of Thompson's group on a CAT(0) boundary. *Groups Geom. Dyn.*, 2(2):185–222, 2008.

[G1] A. Genevois. Coning-off CAT(0) cube complexes. arXiv:1603.06513, 2016.

[G2] A. Genevois. Acylindrical action on the hyperplanes of a CAT(0) cube complex. arXiv:1610.08759, 2016.

[H] F. Haglund. Isometries of CAT(0) cube complexes are semi-simple. available at `https://hal.archives-ouvertes.fr/hal-00148515v1`, 2007.

[L] A. Lonjou. Non simplicité du groupe de Cremona sur tout corps. *Ann. Inst. Fourier (Grenoble)*, 66(5):2021–2046, 2016.

[M1] A. Martin. Acylindrical actions on CAT(0) square complexes. arXiv: 1509.03131, 2015.

[M2] A. Martin. On the acylindrical hyperbolicity of the tame automorphism group of $SL_2(\mathbb{C})$. *Bull. Lond. Math. Soc.*, 49(5):881–894, 2017.

[M3] A. Martin. On the cubical geometry of Higman's group. *Duke. Math. J.* **166** (2017), no. 4, 707–738.

[MO] A. Minasyan and D. Osin. Acylindrical hyperbolicity of groups acting on trees. *Math. Ann.*, 362(3-4):1055–1105, 2015.

[O1] D. V. Osin. Relatively hyperbolic groups: intrinsic geometry, algebraic properties, and algorithmic problems. *Mem. Amer. Math. Soc.*, 179(843):vi+100, 2006.

[O2] D. Osin. Acylindrically hyperbolic groups. *Trans. Amer. Math. Soc.*, 368(2):851–888, 2016.

[S] M. Sageev. CAT(0) cube complexes and groups. In *Geometric group theory*, volume 21 of *IAS/Park City Math. Ser.*, pages 7–54. Amer. Math. Soc., Providence, RI, 2014.

[vdL] H. van der Lek. The homotopy type of complex hyperplane complements. Ph.D. Thesis, University of Nijmegen, 1983.

8

Immutability is not uniformly decidable in hyperbolic groups

Daniel Groves[1]
Department of Mathematics, Statistics, and Computer Science
University of Illinois at Chicago
322 Science and Engineering Offices (M/C 249)
851 S. Morgan St.
Chicago, IL 60607-7045
United States

Henry Wilton[2]
Centre for Mathematical Sciences
University of Cambridge
Wilberforce Road
Cambridge CB3 0WB
United Kingdom

Abstract

A finitely generated subgroup H of a torsion-free hyperbolic group G is called *immutable* if there are only finitely many conjugacy classes of injections of H into G. We show that there is no uniform algorithm to recognize immutability, answering a uniform version of a question asked by the authors.

In [GW1] we introduced the following notion which is important for the study of conjugacy classes of solutions to equations and inequations over torsion-free hyperbolic groups, and also for the study of limit groups over (torsion-free) hyperbolic groups.

[1] The work of the first author was supported by the National Science Foundation and by a grant from the Simons Foundation (#342049 to Daniel Groves).
[2] The second author is partially funded by EPSRC Standard Grant number EP/L026481/1. This paper was completed while the second author was participating in the *Non-positive curvature, group actions and cohomology* programme at the Isaac Newton Institute, funded by EPSRC Grant number EP/K032208/1.

Definition 8.1.1 [GW1, Definition 7.1] Let G be a group. A finitely generated subgroup H of G is called *immutable* if there are finitely many injective homomorphisms $\phi_1, \ldots, \phi_N \colon H \to G$ so that any injective homomorphism $\phi \colon H \to G$ is conjugate to one of the ϕ_i.

We gave the following characterization of immutable subgroups.

Lemma 8.1.2 *[GW1, Lemma 7.2] Let Γ be a torsion-free hyperbolic group. A finitely generated subgroup of Γ is immutable if and only if it does not admit a nontrivial free splitting or an essential splitting over $\mathbb{Z}$.*

The following corollary is immediate.

Corollary 8.1.3 *Let Γ be a torsion-free hyperbolic group and suppose that H is a finitely generated subgroup. If for every action of H on a simplicial tree with trivial or cyclic edge stabilizers H has a global fixed point, then H is immutable.*

If Γ is a torsion-free hyperbolic group then the immutable subgroups of Γ form some of the essential building blocks of the structure of Γ–limit groups. See [GW1] and [GW2] for more information.

In [GW1, Theorem 1.4] we proved that given a torsion-free hyperbolic group Γ it is possible to recursively enumerate the finite tuples of Γ which generate immutable subgroups. This naturally leads us to ask:

Question 8.1.4 [GW1, Question 7.12] Let Γ be a torsion-free hyperbolic group. Is there an algorithm that takes as input a finite subset S of Γ and decides whether or not the subgroup $\langle S \rangle$ is immutable?

We are not able to answer this question, but we can answer the *uniform* version of this question in the negative, as witnessed by Theorem 8.1.6 below. It is worth remarking that the algorithm from [GW1, Theorem 1.4] *is* uniform, in the sense that one can enumerate pairs (Γ, S) where Γ is a torsion-free hyperbolic group (given by a finite presentation) and S is a finite subset of words in the generators of Γ so that $\langle S \rangle$ is immutable in Γ.

As part of our construction in Theorem 8.1.6, we need a particular kind of hyperbolic group, whose properties are listed in the result below. There are many ways one could build such a group; we sketch one below.

Proposition 8.1.5 *There exists a torsion-free hyperbolic group Γ_0 with two subgroups J, J' so that:*

(i) *Γ_0 is rigid;*

(ii) $\langle J, J' \rangle = \Gamma_0$*;*
(iii) $J = \langle x, y \rangle$ *and* $J' = \langle x', y' \rangle$ *are both free of rank* 2*;*
(iv) J *and* J' *are both quasiconvex in* Γ_0*;*
(v) $\{J, J'\}$ *forms a malnormal family in* Γ_0.

Sketch proof We take $\Gamma_0 = \langle x, y, x', y' \mid r \rangle$, where r is a cyclically reduced word that satisfies three conditions:

(a) r satisfies the $C'(1/12)$ small-cancellation condition;
(b) every word of length 12 in the generators is contained in some cyclic conjugate of r or its inverse;
(c) every subword of r of length at least a third the length of r contains every generator.

Thus defined, Γ_0 satisfies the conclusions of [C, Theorem 4.18], and is therefore hyperbolic with Gromov boundary homeomorphic to the Menger sponge; in particular, $\partial_\infty \Gamma_0$ is connected without local cut points, and so Γ_0 is rigid by Bowditch's Theorem [B]. Setting $J = \langle x, y \rangle$ and $J' = \langle x', y' \rangle$, we see that $\Gamma_0 = \langle J, J' \rangle$ by definition. That J and J' are both free of rank 2 is an immediate consequence of Magnus' *Freiheitssatz* [L, Proposition II.5.1]. The remaining two assertions are standard applications of small-cancellation theory (see, for instance, [L, Chapter V]); we sketch their proofs here.

We first show that J is isometrically embedded in the Cayley graph of Γ_0, which clearly implies item (iv). For a contradiction, let j be the shortest element of J whose length in Γ_0 is shorter than its length in J. Consider a disc diagram D with boundary equal to the concatenation $j.g$, where g is a geodesic in the Cayley graph of Γ_0, and suppose that D has at least one 2-cell. Since j is shortest, there are 2-cells e^+ and e^- adjacent to the beginning and end of j respectively. Suppose that at least a 5/6 of the boundary of $e^\pm$ is contained in the boundary of D. Since g is a geodesic, at most half of the boundary of $e^\pm$ is contained in g; therefore at least 1/3 of the boundary of $e^\pm$ is contained in j. But this contradicts assumption (c), since only two generators appear in j. Therefore, less than 5/6 of the boundaries of $e^\pm$ are contained in ∂D. We now appeal to Greendlinger's Lemma [L, Theorem V.4.5], which guarantees a 2-cell e_0, not equal to $e^\pm$, such that a connected subset p of length more than half the boundary of e_0 is contained in ∂D. Since $e_0 \neq e^\pm$, p represents a subword of either j or g. But the first contradicts assumption (c), and the second contradicts the fact that g is a geodesic. This completes the proof that J and J' are isometrically embedded in the Cayley graph.

The proof of item (v) is similar. We will show that J is malnormal. The proof that J' is malnormal is identical, and the proof that all conjugates of J and J' have trivial intersection is similar. Suppose therefore that $j \in J \smallsetminus 1$, $g \in G$ and $j^g = j' \in J$. Then we can build an annular diagram A as in [L, §V.5], with one boundary component equal to j and one boundary component equal to j'. Since no disc of positive area bounds an element of J, we see that A really is homeomorphic to an annulus. Since the boundary is geodesic, each 2-cell in A has at most one interval in each boundary component of A. By assumption (c), those intervals are each of length less than 1/3 the length of r. Combining these facts with the $C'(1/12)$ hypothesis, a standard estimate of the Euler characteristic leads to a contradiction.

This completes our sketch of the proof that $\{J, J'\}$ is malnormal, and therefore of the existence of Γ_0. □

With Γ_0 in hand, we are ready to proceed with the proof of the main theorem.

Theorem 8.1.6 *There is no algorithm that takes a presentation of a (torsion-free) hyperbolic group and a finite tuple of elements as input and determines whether or not the tuple generates an immutable subgroup.*

Proof Let Γ_0 be the group constructed in Proposition 8.1.5 above, along with the free quasiconvex subgroups $J = \langle x, y\rangle$ and $J' = \langle x', y'\rangle$ forming a malnormal family. The group Γ_0, and the subgroups J, J' (with their generating sets) will be fixed throughout the proof.

Consider a finitely presented group Q with unsolvable word problem (see [N]), and let G be a hyperbolic group that fits into a short exact sequence

$$1 \to N \to G \to Q * \mathbb{Z} \to 1,$$

where N is finitely generated and has Kazhdan's Property (T). Such a G can be constructed using [BO, Corollary 1.2], by taking H from that result to be a non-elementary hyperbolic group with Property (T), and recalling that having Property (T) is closed under taking quotients.

Let t be the generator for the second free factor in $Q * \mathbb{Z}$. Given a word u in the generators of Q, define words

$$c_u = tut^{-2}ut,$$

and

$$d_u = utut^{-1}u.$$

Claim 8.1.7 *If $u =_Q 1$ then $\langle c_u, d_u \rangle = \{1\}$ in $Q * \mathbb{Z}$. If $u \neq_Q 1$ then $\langle c_u, d_u \rangle$ is free of rank 2 in $Q * \mathbb{Z}$.*

Proof of Claim 8.1.7 The first assertion of the claim is obvious, and the second follows from the fact that if u is nontrivial in Q, then any reduced word in $\{c_u, d_u\}^{\pm}$ yields a word in $\{t, u\}^{\pm}$ that is in normal form in the free product $Q * \mathbb{Z}$, and hence is nontrivial in $Q * \mathbb{Z}$. □

We lift the elements $c_u, d_u \in Q * \mathbb{Z}$ to elements $\bar{c}_u, \bar{d}_u \in G$ (defined by the same words in the generators).

Claim 8.1.8 *Given words c_u and d_u, it is possible to algorithmically find words $w_u, x_u, y_u, z_u \in N$ so that $\langle w_u \bar{c}_u x_u, y_u \bar{d}_u z_u \rangle$ is quasi-convex and free of rank 2.*

Proof of Claim 8.1.8 It is well known (see, for example, [AGM, Lemma 4.9]) that in a δ-hyperbolic space a path which is made from concatenating geodesics whose length is much greater than the Gromov product at the concatenation points is a uniform-quality quasi-geodesic, and in particular not a loop.

By considering geodesic words representing $\bar{c}_u$ and $\bar{d}_u$, it is possible to find long words in the generators of N as in the statement of the claim so that any concatenation of $(w_u \bar{c}_u x_u)^{\pm}$ and $(y_u \bar{d}_u z_u)^{\pm}$ is such a quasigeodesic. From this, it follows immediately that the free group $\langle w_u \bar{c}_u x_u, y_u \bar{d}_u z_u \rangle$ is quasi-isometrically embedded and has free image in G. This can be done algorithmically because the word problem in G is (uniformly) solvable, so we can compute geodesic representatives for words and calculate Gromov products. □

Let $g_u = w_u \bar{c}_u x_u$ and $h_u = y_u \bar{d}_u z_u$, and let $J_u = \langle g_u, h_u \rangle$. Note that the image of J_u in $Q * \mathbb{Z}$ is either trivial (if $u =_Q 1$) or free of rank 2 (otherwise). Therefore, if $u =_Q 1$ then $J_u \cap N = J_u$ and otherwise $J_u \cap N = \{1\}$.

We now take a second copy of the group G, denoted by G', and likewise use primes to denote elements and subgroups of G'. Consider the group

$$\Gamma_u = G *_{g_u = x, h_u = y} \Gamma_0 *_{g'_u = x', h'_u = y'} G'.$$

Since $\{J = \langle x, y \rangle, J' = \langle x', y' \rangle\}$ forms a malnormal family of quasiconvex subgroups of Γ_0, $\langle g_u, h_u \rangle$ is quasiconvex in G, and $\langle g'_u, h'_u \rangle$ is quasiconvex in G', the group Γ_u is hyperbolic by the Bestvina–Feighn Combination Theorem [BF].

Let $K_u = \langle N, N' \rangle \leq \Gamma_u$. We remark that a presentation for Γ_u and generators for K_u as a subgroup of Γ_u can be algorithmically computed from the presentation of G and the word u.

Claim 8.1.9 *If $u =_Q 1$ then K_u is immutable. If $u \neq_Q 1$ then K_u splits nontrivially over $\{1\}$ and so is not immutable.*

Proof of Claim 8.1.9 Let $N_u = N \cap J_u$, and consider the action of K_u on the Bass–Serre tree of the splitting of Γ_u given by the defining amalgam.

We observed above that if $u =_Q 1$ then $N_u = J_u$, and that if $u \neq_Q 1$ then $N_u = \{1\}$. Thus, if $u =_Q 1$ then $J_u \leq N$, so $J, J' \leq K_u$, and $\Gamma_0 \leq K_u = \langle N, N' \rangle$, and we see that

$$K_u = N *_{J_u = J} \Gamma_0 *_{J'_u = J'} N',$$

which we claim is immutable. Indeed, suppose that K_u acts on a tree T with trivial or cyclic edge stabilizers. Since Property (T) groups have Property (FA) [W], N and N' must act elliptically on T. Moreover, Γ_0 is rigid, so Γ_0 also fixes a point in T. However, if N and Γ_0 do not share a common fixed point, then their intersection (which is free of rank 2) must fix the edge-path between the fixed point sets, contradicting the assumption that edge stabilizers are trivial or cyclic. Thus Γ_0 and N have a common fixed point, and the same argument shows that $\langle N, \Gamma_0 \rangle$ and N' have a common fixed point. Therefore, there is a global fixed point for the K_u-action on T. It follows from Corollary 8.1.3 that K_u is immutable, as required.

On the other hand, if $u \neq_Q 1$ then $N_u = \{1\}$ and it is clear that in this case

$$K_u = N * N',$$

is a nontrivial free splitting of K_u, as required. □

An algorithm as described in the statement of the theorem would (when given the explicit presentation of Γ_u and the explicit generators for K_u) be able to determine whether or not K_u is immutable. In turn, this would decide the word problem for Q, by Claim 8.1.9. Since this is impossible, there is no such algorithm, and the proof of Theorem 8.1.6 is complete. □

Remark 8.1.10 By taking only a cyclic subgroup to amalgamate in the definition of Γ_u, instead of a free group of rank 2, it is straightforward to see that one cannot decide whether non-immutable subgroups split over $\{1\}$ or over $\{\mathbb{Z}\}$.

Acknowledgements

The authors would like to thank the anonymous referee, who pointed out an error in the first version of this paper.

References

[AGM] Ian Agol, Daniel Groves, and Jason Fox Manning. An alternate proof of Wise's Malnormal Special Quotient Theorem. *Forum Math. Pi*, 4:e1, 54, 2016.

[BO] Igor Belegradek and Denis Osin. Rips construction and Kazhdan property (T). *Groups Geom. Dyn.*, 2(1):1–12, 2008.

[BF] M. Bestvina and M. Feighn. A combination theorem for negatively curved groups. *J. Differential Geom.*, 35(1):85–101, 1992.

[B] Brian H. Bowditch. Cut points and canonical splittings of hyperbolic groups. *Acta Math.*, 180(2):145–186, 1998.

[C] Christophe Champetier. Propriétés statistiques des groupes de présentation finie. *Adv. Math.*, 116(2):197–262, 1995.

[GW1] Daniel Groves and Henry Wilton. Conjugacy classes of solutions to equations and inequations over hyperbolic groups. *J. Topol.*, 3(2):311–332, 2010.

[GW2] Daniel Groves and Henry Wilton. The structure of limit groups over hyperbolic groups (English summary). *Isr. J. Math.* 226(1):119–176, 2018.

[L] Roger C. Lyndon and Paul E. Schupp. *Combinatorial group theory.* Springer-Verlag, Berlin, 1977. Ergebnisse der Mathematik und ihrer Grenzgebiete, Band 89.

[N] P. S. Novikov. *Ob algoritmičeskoĭ nerazrešimosti problemy toždestva slov v teorii grupp.* Trudy Mat. Inst. im. Steklov. no. 44. Izdat. Akad. Nauk SSSR, Moscow, 1955.

[W] Yasuo Watatani. Property T of Kazhdan implies property FA of Serre. *Math. Japon.*, 27(1):97–103, 1982.

9

Sphere systems, standard form, and cores of products of trees

Francesca Iezzi
School of Mathematics
University of Edinburgh
James Clerk Maxwell Building
King's Buildings
Mayfield Road
Edinburgh EH9 3JZ
United Kingdom

Abstract

We introduce the concept of a *standard form* for two embedded maximal sphere systems in the doubled handlebody, and we prove an existence and uniqueness result. In particular, we show that pairs of maximal sphere systems in the doubled handlebody (up to homeomorphism) bijectively correspond to square complexes satisfying a set of properties. This work is a variant on Hatcher's normal form.

9.1 Introduction

Let M_g be the connected sum of g copies of $S^2 \times S^1$; M_g is homeomorphic to the double of the handlebody of genus g. The fundamental group of M_g is the free group of rank g, denoted F_g, and, if $Mod(M_g)$ denotes the group of isotopy classes of self-homeomorphisms of M_g, the natural map $Mod(M_g) \to Out(F_g)$ is surjective with finite kernel. Moreover, elements of the kernel fix all homotopy classes of spheres (proven in [L2] p. 80–81). For this reason, collections of spheres in this class of manifolds have been a significant tool in the study of outer automorphism groups of free groups. We refer to a collection of disjoint pairwise-non-isotopic spheres in M_g as a *sphere system*. An important result is that homotopic sphere systems in M_g are isotopic ([L1] Théorème I).

The idea of using sphere systems in M_g as a tool in the study of $Out(F_g)$ goes back to Whitehead ([W], [St]) and has been further developed by

Hatcher in [Ha]. In the latter Hatcher introduces the *sphere complex* of the manifold M_g, which has been a very useful tool in the study of the groups $Out(F_g)$. Collections of spheres in M_g can also be used to define the free factor complex ([HV]), and the Culler–Vogtmann Outer Space (Appendix of [Ha]).

In [Ha] the author also introduces the concept of a *normal form* of spheres with respect to a given maximal sphere system and he proves an existence result. Hatcher's normal form has been the basis of many of the results concerning the sphere complex — for example, the proof of hyperbolicity [HH].

In this paper, we introduce the concept of a *standard form* for a pair of embedded maximal sphere systems (Σ_1, Σ_2) in M_g. Standard form is a symmetric definition, and is equivalent to reciprocal normal form of Σ_1, Σ_2, with the additional requirement that all complementary components of $\Sigma_1 \cup \Sigma_2$ in M_g are handlebodies. We then show an existence and uniqueness result:

Theorem I *Given a pair of maximal sphere systems* (Σ_1, Σ_2) *in* M_g *there exists a homotopic pair* (Σ'_1, Σ'_2) *in standard form.*

Theorem II *If* (Σ_1, Σ_2) *and* (Σ'_1, Σ'_2) *are two homotopic pairs of maximal sphere systems in* M_g *in standard form, then there is a homeomorphism* $F : M_g \to M_g$, *that induces an inner automorphism of the fundamental group and so that* F *maps the pair* (Σ_1, Σ_2) *to the pair* (Σ'_1, Σ'_2).

Note that Theorem I and Laudenbach's result that homotopic sphere systems in M_g are isotopic imply that any pair of maximal sphere systems in M_g can be isotoped to be in standard form.

Our arguments use combinatorial methods. The main idea is that, given a pair of maximal sphere systems (Σ_1, Σ_2), one can build a dual square complex (a CAT(0) cube complex of dimension 2), in which 0-cells correspond to complementary components of $\Sigma_1 \cup \Sigma_2$, 1-cells correspond to components of $\Sigma_1 \setminus \Sigma_2$ and of $\Sigma_2 \setminus \Sigma_1$, and 2-cells correspond to components of $\Sigma_1 \cap \Sigma_2$. The same square complex can also be obtained by applying the construction described in [G] to the dual trees to $\widetilde{\Sigma}_1$ and $\widetilde{\Sigma}_2$ in the universal cover $\widetilde{M}_g$. This idea has been used in [Ho] to estimate distances in Outer Space.

In this paper (Section 9.5) we describe an inverse to this construction, i.e. we show that square complexes endowed with a set of properties determine pairs of sphere systems in M_g in standard form.

The article is organised as follows.

In Section 9.2 we clarify notation and recall how spheres in M_g and intersection numbers relate to partitions of the space of ends of the universal cover.

In Section 9.3 we introduce the definition of *standard form* for a pair of maximal sphere systems (Σ_1, Σ_2), and we hint at how standard form of (Σ_1, Σ_2) implies some properties of the dual square complex.

In Section 9.4 we introduce a more abstract construction: given two trivalent trees T_1, T_2 endowed with a boundary identification, we construct a *core* $C(T_1, T_2)$ and show that the core satisfies certain properties (properties (1)-(5) on page 204). The core $C(T_1, T_2)$ we define turns out to be the same as the *Guirardel core* of T_1 and T_2, defined in [G]. We give a combinatorial description of this object using partitions of the space of ends. The construction of Section 9.4 gives an alternative way of building the dual square complex to two sphere systems in standard form (as shown in Proposition 2.1 of [Ho] and in Proposition 9.4.11 below).

In Section 9.5 we show that, starting with a square complex endowed with properties (1)-(5) of Section 9.4, we can construct a doubled handlebody with two embedded maximal sphere systems in standard form.

As an application of the constructions of Section 9.4 and Section 9.5, in Section 9.6 we prove Theorem I and Theorem II.

As a note for the reader, it is worth pointing out that the construction described in Section 9.3.1 and in Section 9.3.2 is introduced early on in the paper as a way of conveying intuition, but will not be used to prove Theorem I. The construction will only play a role in the proof of Theorem II.

Throughout the paper, for the sake of simplicity, we always assume that a pair of maximal sphere systems (Σ_1, Σ_2) in M_g satisfies the following hypothesis:

$(*)$ no sphere in Σ_1 is homotopic to any sphere in Σ_2

All the arguments of the article can be generalised to the case where $(*)$ is not fulfilled. A discussion about this more general case and a hint on how to generalise the arguments can be found in Section 2.6 of [I].

Note that Theorem I could also be proved using Hatcher's existence theorem for normal form. However, our arguments are independent of Hatcher's work, even though the concept of normal form has served as an inspiration.

Acknowledgements

This work was carried out during my PhD studies under the supervision of Brian Bowditch. I am very grateful for his guidance. My PhD was funded by an EPSRC doctoral grant. I have written this article while supported by the Warwick Institute of Advanced Study (IAS) and the Warwick Institute of Advanced Teaching and Learning (IATL).

9.2 Spheres, partitions and intersections

Throughout the paper, we denote by M_g the connected sum of g copies of $S^2 \times S^1$ (i.e. the doubled handlebody), and by $\widetilde{M}_g$ its universal cover. We always suppose spheres in M_g are embedded and intersect transversely. A sphere is *essential* if it does not bound a ball; in the paper we will always assume spheres are essential, unless stated otherwise.

We denote by $i(s_1, s_2)$ the minimal possible number of circles belonging to $s_1 \cap s_2$, over the homotopy class of s_1 and s_2, and we call this number the *intersection number* of the spheres s_1 and s_2. We say that two spheres s_1, s_2 *intersect minimally* if they realise their intersection number. A *sphere system* in M_g is a collection of non-isotopic disjoint spheres. We call a sphere system Σ *maximal* if it is maximal under inclusion, i.e. any embedded essential sphere σ is either isotopic to a component of Σ or intersects Σ. A maximal sphere system Σ in M_g contains $3g-3$ spheres, and all connected components of $M_g \smallsetminus \Sigma$ are 3-holed 3-spheres.

This section contains some (already known) background results about embedded spheres in the manifold M_g. In particular, we will recall that embedded spheres in $\widetilde{M}_g$ can be identified with partitions of the space of ends of $\widetilde{M}_g$, which can be identified with the boundary of a given tree (Lemma 9.2.1); furthermore, the intersection number of two spheres in $\widetilde{M}_g$ is positive if and only if the partitions associated to the two spheres satisfy a particular property (Lemma 9.2.2).

We first recall that, given a sphere system Σ in M_g (or in $\widetilde{M}_g$), we can associate to Σ a graph G_Σ. Namely we take a vertex v_C for each component C of $M_g \smallsetminus \Sigma$ and an edge e_σ for each sphere σ in Σ. The edge e_σ is incident to the vertex v_C if the sphere σ is one of the boundary components of C. We call G_Σ the *dual graph* to Σ. We can endow G_Σ with a metric by setting the length of each edge to one. There is a natural retraction $r : M_g \to G_\Sigma$: consider a regular neighbourhood of Σ, call it $U(\Sigma)$ and parametrise it as $\Sigma \times (0,1)$. For any component C of $M_g \smallsetminus U(\Sigma)$, let $r|C$ map everything to the vertex v_C. For any sphere

σ in Σ, set $r(\sigma \times t)$ to be the point t in e_σ. If each complementary component of Σ in M_g is simply connected, then the retraction r induces an isomorphism of fundamental groups. Note that if $\widetilde{\Sigma}$ is the full lift of Σ in $\widetilde{M}_g$, then the dual graph to $\widetilde{\Sigma}$ in $\widetilde{M}_g$ (which is a tree) is isomorphic to the universal cover of the graph G_Σ. We will denote it by T_Σ and call it the *dual tree* to M_g and Σ (or the dual tree to $\widetilde{M}_g$ and $\widetilde{\Sigma}$). Note also that the retraction $r : M_g \to G_\Sigma$ lifts to a retraction $h : \widetilde{M}_g \to T_\Sigma$. If Σ is a maximal sphere system, then the dual graph G_Σ is trivalent, as is the dual tree T_Σ.

9.2.1 Space of ends

Next we recall the definition of the space of ends. Let X be a topological space and let $\{K_n\}$ be any exhaustion of X by compact sets. Consider all sequences $\{U_n\}$ where U_k is a component of $X \setminus K_k$ and $U_k \supset U_{k+1}$, under the following equivalence relation: the sequence $\{U_n\}$ is equivalent to the sequence $\{V_n\}$ if for each U_i there exists j with $V_j \subset U_i$ and for each V_j there exists i with $U_i \subset V_j$. Then an *end* of X is an equivalence class of sequences $\{U_n\}$. One can check that this definition does not depend on the particular sequence of compact sets we choose. Given an open set A in X we say that an end $\{U_n\}$ is contained in the set A if, for k large enough, the set U_k is contained in A. Call the collection of ends of a given space X the *space of ends* of X and denote it by $End(X)$. The space $End(X)$ can be endowed with a topology: a fundamental system of neighbourhoods for the end $\{U_n\}$ is given by the sets $\{e_{U_k}\}$, for $U_k \in \{U_n\}$, where e_{U_k} consists of all the points in $End(X)$ contained in U_k.

Note that the space of ends of a tree can be identified with its Gromov boundary, which is a Cantor set. We refer to chapter 8 of [BH] and to [P] for some more detailed background.

We observe next that the space of ends of the manifold $\widetilde{M}_g$ can be identified with the space of ends of a given tree. In fact, consider M_g with an embedded maximal sphere system Σ and let T_Σ be the dual tree to M_g and Σ (note that T_Σ is trivalent by maximality of Σ). The retraction $h : \widetilde{M}_g \to T_\Sigma$ induces a homeomorphism between the space $End(T_\Sigma)$ and the space $End(\widetilde{M}_g)$. The latter is therefore a Cantor set.

Now, since $\widetilde{M}_g$ is simply connected, every sphere $\sigma \subset \widetilde{M}_g$ separates, and induces a partition of the space of ends. The partition induced by a sphere σ on the space of ends of $\widetilde{M}_g$ coincides with the partition induced by the corresponding edge e_σ on the boundary of the dual tree T_Σ.

To fix terminology, if C is a Cantor set and $P_1 \doteq C = (P_1^+ \cup P_1^-)$, $P_2 \doteq C = (P_2^+ \cup P_2^-)$ are two distinct partitions of the set C, we say that P_1 and P_2 are *non-nested* if all four sets $P_1^+ \cap P_2^+$, $P_1^+ \cap P_2^-$, $P_1^- \cap P_2^+$, $P_1^- \cap P_2^-$ are non-empty. We say that P_1 and P_2 are *nested* otherwise. The following holds:

Lemma 9.2.1 *For any clopen partition P of $End(\widetilde{M}_g)$ there is a sphere s_P embedded in $\widetilde{M}_g$ that induces the partition P. Two embedded spheres are homotopic if and only if they induce the same partition on $End(\widetilde{M}_g)$.*

Now recall that we say two spheres, s_1 and s_2, intersect minimally if the number of components of $s_1 \cap s_2$ is minimal over the homotopy classes of s_1 and s_2. Then the following holds:

Lemma 9.2.2 *Two non-homotopic embedded minimally-intersecting spheres s_1, s_2 in $\widetilde{M}_g$ intersect at most once, and they intersect if and only if the partitions induced by s_1 and s_2 on the space of ends of $\widetilde{M}_g$ are non-nested.*

We refer to Section 2.1.1 of [I] (in particular Claim 2.1.6, Lemma 2.1.7, Lemma 2.1.9, Lemma 2.1.10) for proofs of Lemma 9.2.1 and Lemma 9.2.2 above. Note that such proofs do not rely on Hatcher's normal form.

9.3 Standard form for sphere systems, piece decomposition and dual square complexes

In this section we introduce the definition of a *standard form* for sphere systems, and we describe some properties of this standard form. Standard form is a refinement of Hatcher's normal form. Loosely speaking, two embedded maximal sphere systems in M_g are in standard form if they intersect minimally and, in addition, their complementary components are as simple as possible.

To clarify terminology, if Σ_1 and Σ_2 are two sphere systems in M_g and $\widetilde{\Sigma}_1$, $\widetilde{\Sigma}_2$ are their full lifts to the universal cover $\widetilde{M}_g$, we say that Σ_1 and Σ_2 are in *minimal form* if each sphere in $\widetilde{\Sigma}_1$ intersects each sphere in $\widetilde{\Sigma}_2$ minimally.

A priori, our definition of minimal form seems stronger than the most intuitive definition (requiring the number of components of $\Sigma_1 \cap \Sigma_2$ to be minimal over the homotopy class of Σ_1 and Σ_2). Indeed, both Hatcher's work [Ha], and Theorem 9.6.1 below imply that the two definitions of minimality are equivalent; furthermore, if one of the two systems is maximal, then minimal form is equivalent to Hatcher's normal form (Lemma 7.2 in [HOP]). However, our arguments will not use the equivalence of these definitions, and will be independent of Hatcher's work.

Definition 9.3.1 Let Σ_1 and Σ_2 be two embedded maximal sphere systems in M_g. We say that Σ_1 and Σ_2 are in *standard form* if they are in minimal form with respect to each other and, in addition, all the complementary components of $\Sigma_1 \cup \Sigma_2$ in M_g are handlebodies. We define the standard form for two maximal sphere systems $\widetilde{\Sigma}_1$ and $\widetilde{\Sigma}_2$ in the universal cover $\widetilde{M}_g$ in the same way.

The existence of a standard form for any two maximal sphere systems can be deduced from Proposition 1.1 of [Ha]; we give an alternative proof below (Theorem 9.6.1). We also prove a sort of uniqueness for standard form (Theorem 9.6.2).

In Section 9.3.1 and Section 9.3.2, we show how to associate a dual square complex to a pair of maximal sphere systems in standard form. However such arguments will not be used to prove existence of standard form (Theorem 9.6.1), but only to prove uniqueness (Theorem 9.6.2).

9.3.1 Piece decomposition

Given the manifold M_g and two embedded maximal sphere systems Σ_1 and Σ_2 in standard form, we colour Σ_1 with black and Σ_2 with red. We will call the components of $M_g \setminus (\Sigma_1 \cup \Sigma_2)$ the *3-pieces* of $(M_g, \Sigma_1, \Sigma_2)$, we call the components of $\Sigma_1 \setminus \Sigma_2$ the *2-pieces of* Σ_1, or the *black 2-pieces*, and the components of $\Sigma_2 \setminus \Sigma_1$ the *2-pieces of* Σ_2, or the *red 2-pieces*. Finally, we call the components of $\Sigma_1 \cap \Sigma_2$ the *1-pieces* of $(M_g, \Sigma_1, \Sigma_2)$. The manifold M_g is the union of 1-pieces, 2-pieces and 3-pieces. We call this collection of pieces a *piece decomposition* for the triple $(M_g, \Sigma_1, \Sigma_2)$. In the same way, we can define a piece decomposition for the triple $(\widetilde{M}_g, \widetilde{\Sigma}_1, \widetilde{\Sigma}_2)$. Note that, since the complementary components of maximal sphere systems are simply connected, pieces of $(M_g, \Sigma_1, \Sigma_2)$ lift homeomorphically to pieces of $(\widetilde{M}_g, \widetilde{\Sigma}_1, \widetilde{\Sigma}_2)$; hence, to study the properties of a piece decomposition, we can analyse pieces of $(M_g, \Sigma_1, \Sigma_2)$ or pieces of $(\widetilde{M}_g, \widetilde{\Sigma}_1, \widetilde{\Sigma}_2)$, according to what is most convenient. We go on to describe some features of a piece decomposition for a triple $(M_g, \Sigma_1, \Sigma_2)$.

Recall that, by maximality, all the components of $M_g \setminus \Sigma_1$ and $M_g \setminus \Sigma_2$ are 3-holed 3-spheres, and that we are assuming hypothesis $(*)$ in the introduction.

As for 1-pieces, since Σ_1 and Σ_2 intersect transversely, all 1-pieces of $(M_g, \Sigma_1, \Sigma_2)$ are circles.

Now, let p be a 2-piece of $\widetilde{\Sigma}_2$. First, p is a planar surface (for p is a subsurface of a sphere $\sigma \in \widetilde{\Sigma}_2$, and moreover, by maximality and hypothesis $(*)$, $\sigma \cap \widetilde{\Sigma}_1$ is non-empty). Further, p is contained in a component of $\widetilde{M}_g \smallsetminus \widetilde{\Sigma}_1$ (which is a 3-holed 3-sphere by maximality of $\widetilde{\Sigma}_1$), and, by minimal form and Lemma 9.2.2, p cannot intersect the same sphere of $\widetilde{\Sigma}_1$ in more than one circle. This implies that p has at most three boundary components. Summarising, a red 2-piece p embedded in a component C of $\widetilde{M}_g \smallsetminus \widetilde{\Sigma}_1$ is either a disc, or an annulus, or a pair of pants. If p is a disc, then, by minimal form, ∂p lies on a boundary component of C and p separates the other two components. If p is an annulus or a pair of pants, then different components of ∂p lie on different components of ∂C. The same requirements hold for black 2-pieces in $\widetilde{M}_g$ (by symmetry), and for 2-pieces of $(M_g, \Sigma_1, \Sigma_2)$.

Note that these are exactly the conditions required by Hatcher's normal form.

3-pieces are all handlebodies, by definition of standard form. Further, the boundary of a 3-piece is the union of black 2-pieces, red 2-pieces, and 1-pieces that are adjacent to a black 2-piece and a red 2-piece. Given a 3-piece P, we use the term *boundary pattern* for P to refer to the union of 2-pieces and 1-pieces composing ∂P. Note that, since the complementary components of maximal sphere systems are 3-holed 3-spheres, the boundary pattern of a 3-piece contains at most three black 2-pieces and three red 2-pieces. Indeed, standard form imposes further conditions on boundary patterns and it turns out that there are only nine possibilities, which are drawn, from different perspectives, on the left hand sides of Figure 9.1 and Figure 9.5. In particular, the genus of a 3-piece is at most four. We omit a proof of this here, since it is not essential to the arguments, and refer to Appendix B of [I].

In the remainder we will often need to work with the closure of 2-pieces and 3-pieces. Therefore, with a little abuse of terminology, we will sometimes use the terms "2-piece" and "3-piece" also when referring to the closure of the pieces defined above.

9.3.2 Dual square complexes

Given a piece decomposition for the triple $(M_g, \Sigma_1, \Sigma_2)$, we can construct a dual square complex. We will show later that such a square complex satisfies some very special properties.

Digression on square complexes

We make a short digression and recall some basic definitions and facts concerning square complexes.

Recall that a square complex is a cube complex of dimension 2, i.e, 1-cells are unit intervals and 2-cells are unit euclidean squares. Each 2-cell is attached along a loop of four 1-cells. We can endow a square complex with a path metric by identifying each 1-cell to the unit interval and each 2-cell to the euclidean square $[0,1] \times [0,1]$. In the remainder we will use the word "vertex"to refer to a 0-cell, the word "edge"to refer to a 1-cell and the word "square"to refer to a 2-cell. A square complex is said to be V-H (Vertical-Horizontal) if each edge can be labelled as vertical or horizontal, and vertical and horizontal 1-cells alternate on the attaching loop of each square.

An important concept is that of a "hyperplane", which we shortly define below. First we define a *midsquare* as a unit interval contained in a square, parallel to one of the edges, containing the barycentre of the square. Consider now the equivalence relation $\sim$ on the edges of a square complex Δ generated by $e \sim e'$ if e and e' are opposite edges of the same square in Δ. Given an equivalence class of edges $[e]$ in Δ we define the *hyperplane* dual to $[e]$ as the set of midsquares in Δ intersecting edges in $[e]$. Note that hyperplanes of a square complex are connected graphs, and that if Δ is a CAT(0) square complex, then two hyperplanes in Δ intersect at most once.

As a last note, recall that, by a generalisation of the Cartan–Hadamard Theorem ([BH] p. 193), a simply connected locally-CAT(0) metric space is CAT(0) (where the term locally-CAT(0) means that each point has a CAT(0) neighbourhood). We refer to [Sa] for some background on cube complexes and to [BH] for a more detailed discussion on CAT(0) metric spaces.

Now, given a triple $(M_g, \Sigma_1, \Sigma_2)$ with Σ_1, Σ_2 in standard form, we can naturally construct a square complex: vertices correspond to 3-pieces, edges correspond to 2-pieces and squares correspond to 1-pieces. For $n = 1, 2$, an n-cell is attached to an $(n-1)$-cell if the piece corresponding to the former lies on the boundary of the piece corresponding to the latter. We denote this complex by $\Delta(M_g, \Sigma_1, \Sigma_2)$ and call it the square complex *dual* to Σ_1 and Σ_2 in M_g. We colour with black the edges corresponding to 2-pieces of Σ_1 and with red the edges corresponding to 2-pieces of Σ_2. In the same way we can construct a square complex $\Delta(\widetilde{M}_g, \widetilde{\Sigma}_1, \widetilde{\Sigma}_2)$, dual to the triple $(\widetilde{M}_g, \widetilde{\Sigma}_1, \widetilde{\Sigma}_2)$.

Note that the action of the free group F_g on the manifold $\widetilde{M}_g$ induces an action of the same group on the complex $\Delta(\widetilde{M}_g, \widetilde{\Sigma}_1, \widetilde{\Sigma}_2)$ and the quotient of this square complex under the group action is the square complex $\Delta(M_g, \Sigma_1, \Sigma_2)$.

When no ambiguity can occur we denote the complex $\Delta(M_g, \Sigma_1, \Sigma_2)$ by Δ, and the complex $\Delta(\widetilde{M}_g, \widetilde{\Sigma}_1, \widetilde{\Sigma}_2)$ by $\widetilde{\Delta}$.

Digression: properties of the dual square complex

We describe now some properties the complex $\widetilde{\Delta}$ must satisfy (Lemma 9.3.2–Lemma 9.3.6), and we hint at some of the proofs. The reader may skip this part, since Lemma 9.3.2–Lemma 9.3.6 will be restated and proved in the next section, after we introduce a more abstract way of constructing $\widetilde{\Delta}$.

Lemma 9.3.2 *$\widetilde{\Delta}$ is a connected, locally finite, V-H square complex.*

We will treat the black edges as horizontal and the red ones as vertical. We will call the segments parallel to vertical (resp. horizontal) edges *vertical* (resp. *horizontal*) *lines.*

Lemma 9.3.3 *The complex $\widetilde{\Delta}$ is endowed with two surjective projections, p_1, p_2, onto two infinite trivalent trees, T_1 and T_2. The projection p_1 (resp. p_2) corresponds to collapsing the vertical (resp. horizontal) lines to points.*

Indeed, the trees T_1 and T_2 correspond respectively to the dual trees to $\widetilde{\Sigma}_1$ and $\widetilde{\Sigma}_2$.

Lemma 9.3.4 *Hyperplanes in $\widetilde{\Delta}$ are finite trees.*

Indeed, hyperplanes dual to red (resp. black) edges correspond to spheres in $\widetilde{\Sigma}_1$ (resp. $\widetilde{\Sigma}_2$). Edges and vertices of a hyperplane correspond respectively to 1-pieces and 2-pieces contained in the corresponding sphere.

As for a vertex link in $\widetilde{\Delta}$, this is entirely determined by the boundary pattern of the 3-piece the vertex corresponds to. Such a link is necessarily a subgraph of the bipartite graph $K_{(3,3)}$, since a boundary pattern contains at most three red 2-pieces and three black 2-pieces, where two 2-pieces of the same colour cannot be adjacent. Restrictions imposed by standard form allow us to list all possible boundary patterns and corresponding vertex links (which are described in Figure 9.1). In fact, the following holds:

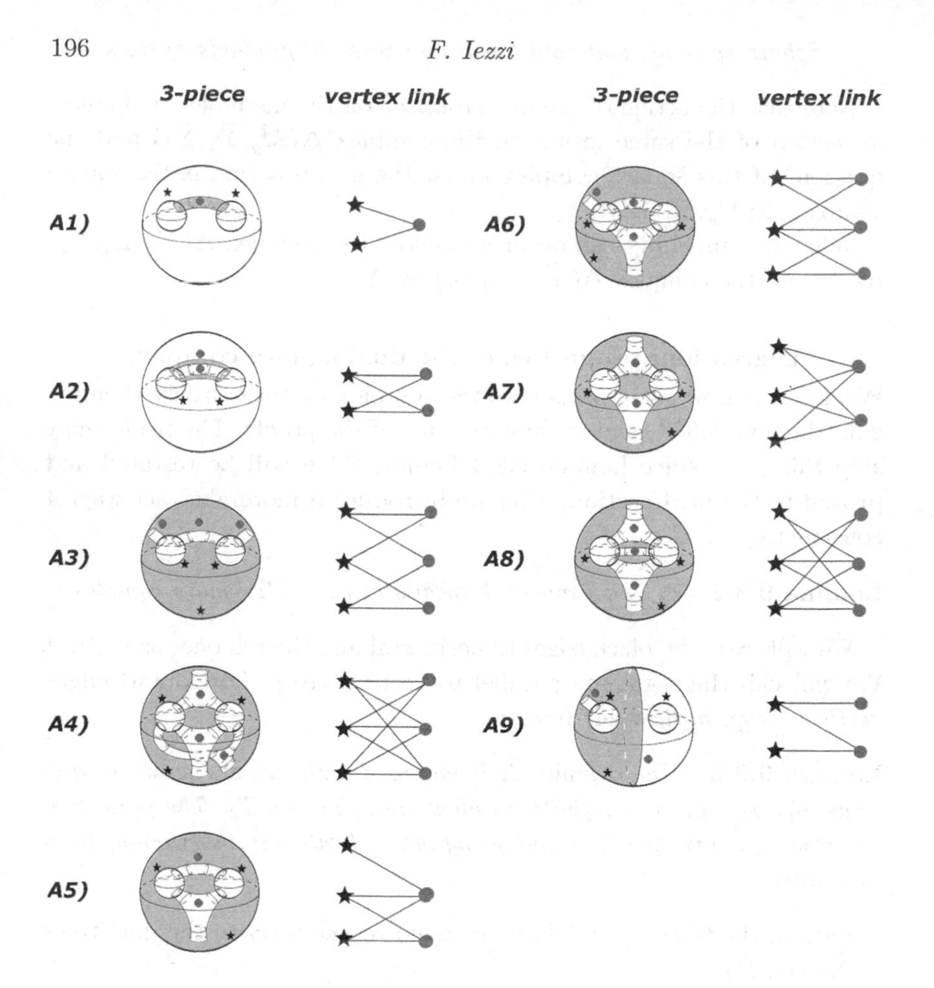

Figure 9.1 All the possibilities for 3-pieces in M_g with the corresponding vertex links. The 3-piece we consider is the part outlined with grey. Pictures are drawn in one dimension lower, i.e. we draw a section of each piece, for example a circle represents a sphere, two parallel lines represent an annulus, etc. In each picture the three black circles represent three spheres of Σ_1 bounding a component C of $M_g \setminus \Sigma_1$. These are labelled in each picture with black stars, corresponding to the black starred vertices in the corresponding vertex-link. The lines labelled with dots represent the 2-pieces of Σ_2 in a complementary component of Σ_1, corresponding to the dotted vertices.

Lemma 9.3.5 *All the possible vertex links in $\widetilde{\Delta}$ are the nine graphs listed in Figure 9.1.*

Another important property is the following:

Lemma 9.3.6 *The complex $\Delta(\widetilde{M}_g, \widetilde{\Sigma}_1, \widetilde{\Sigma}_2)$ is simply connected.*

Note that simply connectedness of the complex $\widetilde{\Delta}$ implies that $\widetilde{\Delta}$ is the universal cover of the complex Δ and, as a consequence, the latter complex has the free group F_g as its fundamental group.

9.4 The core of two trees

In this section we describe an abstract construction. Starting with two trees endowed with geometric actions by the free group F_g, we construct a square complex C and call it the *core* of the product of the two trees.

We show later (Proposition 9.4.11) that the method described below provides an alternative way of constructing the square complex dual to two maximal sphere systems in standard form in $\widetilde{M}_g$.

Let T and T' be two three-valent trees both endowed with a free, properly discontinuous and cocompact action by the free group F_g; call these actions ρ and ρ'.

Note that the action ρ (resp. ρ') induces a canonical identification of the boundary of the group F_g to the Gromov boundary of the tree T (resp. T'). Hence the boundaries of T and T' can also be identified. In view of this identification, in the remainder we will often use the term *boundary space* to refer to the Gromov boundary of T and the Gromov boundary of T'.

Each edge in T or T' induces a partition on the boundary space. If e is an edge in T, we will denote by P_e the partition induced by e on the boundary space and by e^+ and e^- the two sets composing this partition. If e is an edge in T and e' is an edge in T', we say that the induced partitions P_e and $P_{e'}$ are *non-nested* if no set of one partition is entirely contained in a set of the other partition, namely all sets $e^+ \cap e'^+$, $e^+ \cap e'^-$, $e^- \cap e'^+$, $e^- \cap e'^-$ are non-empty. We say that P_e and $P_{e'}$ are *nested* otherwise.

In the remainder we will assume the following hypothesis:

$(**)$ There do not exist an edge e in T and an edge e' in T' inducing the same partition on the boundary space.

It will be clear in the next section that hypothesis $(**)$ above corresponds to hypothesis $(*)$ in the introduction. The construction described below can be generalised to the case where hypothesis $(**)$ is not fulfilled. We refer to Section 2.6.3 of [I] for a discussion of this more general case.

Consider now the product $T \times T'$: it is a CAT(0) square complex, where each vertex link is the bipartite graph $K_{3,3}$. This space can be naturally endowed with a diagonal action γ. Namely, given a vertex (v_1, v_2) in $T \times T'$

and an element g of F_g, we set $\gamma_g(v_1, v_2)$ to be the vertex $(\rho_g(v_1), \rho'_g(v_2))$. Since ρ and ρ' are free and properly discontinuous, so is γ.

We go on to define the main object of this section.

Definition 9.4.1 The *core* of $T \times T'$ is the subcomplex of $T \times T'$ consisting of all the squares $e \times e'$ such that e is an edge in T, e' is an edge in T', and the two partitions induced by e and e' on the boundaries of T and T' are non-nested. We will denote this complex by $C(T,T')$. Where no ambiguity can occur, we will write C instead of $C(T,T')$.

The complex C is invariant under the diagonal action of the group F_g. In fact, for each g in F_g the maps ρ_g and ρ'_g induce the same homeomorphism on $\partial T = \partial T'$, therefore the partitions induced by the edges e and e' are nested if and only if those induced by the edges $\rho_g(e)$ and $\rho'_g(e')$ are. Denote by $\Delta(T,T')$ the quotient of $C(T,T')$ by the diagonal action of the group F_g.

Remark 9.4.2 Note that in order to define the complex $C(T,T')$ we do not need the group actions, but only an identification of ∂T and $\partial T'$. Furthermore, the above construction can be applied to any pair of simplicial trees.

Remark 9.4.3 It is not difficult to check that the core $C(T,T')$ defined above actually coincides with the Guirardel core of the trees T and T', introduced in [G] (we refer to [G] or to Section 1 of [BBC] for the exact definition). This equivalence can also be deduced from [BBC]. Hence, the construction described above gives a combinatorial way of defining the Guirardel core of any two simplicial trees. However Guirardel's construction is much more general, since it can also be applied to $\mathbb{R}$-trees.

9.4.1 Properties of the core

This section is aimed at showing that the complex $C(T,T')$ satisfies Lemmas 9.3.2–9.3.6.

First note that, as a subcomplex of $T \times T'$, the complex $C(T,T')$ is a V-H square complex, and is naturally endowed with two projections: $\pi_T : C \to T$ and $\pi_{T'} : C \to T'$. If e is an edge in T (resp. e' is an edge in T'), we use the term *preimage* of e (resp. of e'), to denote the preimage of the edge e under the map $\pi_T : C \to T$ (resp. of the edge e' under the map $\pi_{T'} : C \to T'$). In the same way we define the preimages of vertices. The next goal is to prove the following:

Proposition 9.4.4 *The core $C(T,T')$ is a connected square complex and the quotient $\Delta(T,T')$ is finite.*

Note that Proposition 9.4.4 can be deduced from [G], using the fact that the $C(T,T')$ coincides with the Guirardel core of T and T'. For the sake of completeness, we give below an independent proof using mostly combinatorial methods. The proof will consist of several steps.

To establish terminology, given an edge $e \in T$, we denote by T'_e the subset of T' consisting of edges e' in T' such that the partitions induced by e and e' are not nested. Note that T'_e coincides with the hyperplane dual to e and that the preimage of an edge e is exactly the interval bundle over T'_e. We will prove the following:

Lemma 9.4.5 *For each edge $e \in T$, T'_e is a finite subtree of T'.*

Proof By definition, given $e \in T$, the edge e' belongs to T'_e if and only if all the sets $e'^+ \cap e^+$, $e'^- \cap e^+$, $e'^+ \cap e^-$ and $e'^- \cap e^-$ are non-empty. We first prove that T'_e is connected by showing that if two edges a and b in T' belong to T'_e, then the geodesic in T' joining a and b is contained in T'_e. To prove this, we may suppose $a^+ \supset c^+ \supset b^+$, and as a consequence $a^- \subset c^- \subset b^-$. Now, since the sets $b^+ \cap e^+$ and $b^+ \cap e^-$ are non-empty, the sets $c^+ \cap e^+$ and $c^+ \cap e^-$ are respectively non-empty; since the sets $a^- \cap e^+$ and $a^- \cap e^-$ are non-empty, then the sets $c^- \cap e^+$ and $c^- \cap e^-$ are also non-empty. Hence, the edge c belongs to T'_e.

Next, we prove that T'_e is finite. By connectedness of T'_e, it is sufficient to show that if $r' = \{e'_i\}$, with $i \in \mathbb{N}$, is a geodesic ray in T', then the subset of r' contained in T'_e is finite, i.e. there exists $I \in \mathbb{N}$ such that the set e'^+_i (or e'^-_i) is contained in one of the two sets e^+ or e^- for each $i \geq I$. To prove that suppose $e'^+_{i+1} \subset e'^+_i$, and note that the limit of the ray $\{e'_i\}$ is a point p in $\partial T'$ — we may assume p belongs to e^+. Now suppose that both sets $e'^+_i \cap e^+$ and $e'^+_i \cap e^-$ are non-empty for each i. Then, since these are compact subsets of the compact space $\partial T'$, both sets $(\cap_{i \in \mathbb{N}} e'^+_i) \cap e^+$ and $(\cap_{i \in \mathbb{N}} e'^+_i) \cap e^-$ are non-empty, which leads to a contradiction, since $(\cap_{i \in \mathbb{N}} e'^+_i) = p \in e^+$. □

Remark 9.4.6 As an immediate consequence of Lemma 9.4.5, all the preimages of edges in T (and, by symmetry, all the preimages of edges $e' \in T'$) are finite and connected, since they are interval bundles over finite trees; moreover hyperplanes in $C(T,T')$ are finite trees.

As another consequence of Lemma 9.4.5 we can prove the following:

Proposition 9.4.7 *The quotient space $\Delta(T,T')$ is a finite V-H square complex, and hyperplanes in $\Delta(T,T')$ are trees.*

Proof As usual, denote the actions of F_g on T and T' by ρ and ρ' respectively, and the diagonal action of F_g on $T\times T'$ by γ. Note that, by invariance of the core, if e is any edge in T, g is any element of F_g, and F_e is the preimage of the edge e in $C(T,T')$, then the preimage $F_{\rho_g(e)}$ of the edge $\rho_g(e)$ is exactly $\gamma_g(F_e)$. Note also that T/F_g is a finite graph, by cocompactness of the action ρ. Hence finiteness of edge preimages immediately implies that the complex $\Delta(T,T')$ is finite.

Now, $\Delta(T,T')$ is a V-H square complex; since $C(T,T')$ is, and the diagonal action γ maps vertical (resp. horizontal) edges to vertical (resp. horizontal) edges.

To prove that hyperplanes in $\Delta(T,T')$ are trees, we observe that, since the actions ρ and ρ' are free, two squares belonging to the preimage of the same edge cannot be identified under the quotient map. Therefore the restriction of the quotient map to a hyperplane is a graph isomorphism. □

To prove connectedness of the core $C(T,T')$ we still need some preliminary lemmas. While proving the next lemma, we use hypothesis $(**)$, i.e. we suppose that there do not exist an edge in T and an edge in T' inducing the same partition on ∂T.

Lemma 9.4.8 *The projections $\pi_T : C\to T$ and $\pi_{T'} : C\to T'$ are both surjective.*

Proof We prove that π_T is surjective, i.e. for each edge e in T there exists an edge e' in T' such that $e\times e'$ is in C or equivalently such that the partition induced by e and the one induced by e' are non-nested. The same argument can be used to prove that the projection $\pi_{T'}$ is surjective. Let e be any edge in T. As usual we denote by $P_e = e^+\cup e^-$ the partition induced by the edge e.

First we claim that there are edges a and b in T' such that $a^+\subset e^+\subset b^+$. To prove the claim note first that, since the preimage F_e is finite by Remark 9.4.6, there exists at least one edge a in T' such that the partitions P_e and P_a are nested. We may suppose without loss of generality that $a^+\subset e^+$. Now, pick a point p in e^- and let $r = a, e'_1, e'_2\dots$ be the geodesic ray in T' joining the edge a to the point p. Since we have $e^-\subset a^-$, the point p belongs to a^-. Consequently we have the containment $e'^+_i\subset e'^+_{i+1}$ for each e_i' in the geodesic ray r. The set $\bigcup_i e'^+_i$ coincides with $\partial T'\setminus p$, and therefore it contains e^+. Since e^+ is compact, there exists a natural number I such that e'^{+}_{i} contains e^+ for each i greater than or equal to I, which proves the claim.

We may as well choose I to be minimal in the set $\{i : e'^{+}_{i} \supset e^+\}$, and denote e'_I by b.

To conclude the proof of Lemma 9.4.8, we will now show that one of the two edges adjacent to b in T' is contained in $\pi_T^{-1}(e)$.

Let r be the geodesic defined above and denote by c the edge immediately preceding b on the ray r. We know that $a^+ \subset c^+ \subset b^+$ (equivalently $a^- \supset c^- \supset b^-$), that $b^+ \supset e^+ \supset a^+$, and that c^+ does not contain e^+ (this follows from the fact that b is the first edge in the ray r such that b^+ contains e^+). Recall that, by hypothesis $(**)$, no containment can be an equality. There are then two possibilities.

–Case 1: $c^+ \not\subset e^+$. In this case the partition induced by e and the one induced by c are non-nested. In fact, $c^+ \not\supset e^+$ implies $e^+ \cap c^-$ is non-empty, and $c^+ \not\subset e^+$ implies $c^+ \cap e^-$ is non-empty. In addition, $e^+ \cap c^+$ is non-empty since $a^+ \subset e^+$ and $a^+ \subset c^+$, and $e^- \cap c^-$ is non-empty since $b^- \subset e^-$ and $b^- \subset c^-$.

–Case 2: $c^+ \subset e^+$. In this case let d denote edge in T' adjacent to both b and c and denote by v the vertex in T' where the edges c, b, d intersect. We claim that the partitions induced by d and e are not nested. To prove the claim, first observe that the vertex v induces a partition $\partial T' = D_1 \cup D_2 \cup D_3$ where D_1 is equal to c^+, D_2 is equal to b^- and D_3 is equal to $\partial T' \setminus (D_1 \cup D_2)$; hence the partition induced by d is $\partial T' = (D_1 \cup D_2) \cup D_3$. Now, since e^+ strictly contains $c^+ = D_1$ and is strictly contained in $D_1 \cup D_3 = b^+$, both sets $e^+ \cap D_3$ and $e^- \cap D_3$ are non-empty. Moreover, since D_1 is contained in e^+ the set $(D_1 \cup D_2) \cap e^+$ is non-empty, and since e^+ is contained in $b^+ = D_1 \cup D_3$, it follows that D_2 is contained in e^- and therefore the set $(D_1 \cup D_2) \cap e^-$ is non-empty. Consequently the partitions induced by e and d are non-nested, which concludes the proof. □

We have already shown that the preimage of each edge through the projections π_T and $\pi_{T'}$ is connected. The following lemma will allow us to conclude the proof of the connectedness of the core.

Lemma 9.4.9 *The preimage of each vertex is a finite tree, in particular it is connected.*

Proof Let $v \in T$ be a vertex, and denote by F_v its preimage through $\pi_T : C \to T$. We show that F_v is a finite tree. By symmetry, the same arguments show that for any vertex $v' \in T'$, its preimage through $\pi_{T'}$ is a finite tree.

Note first that if we denote the three edges incident to the vertex v by e_1, e_2 and e_3, then the preimage F_v is the union $T'_{e_1} \cup T'_{e_2} \cup T'_{e_3}$ (recall that T'_{e_i} is the subtree of T' consisting of all the edges e' such that the partitions induced by e' and e_i are non-nested). Hence F_v is the union of three finite trees; we go on to prove that F_v is connected.

To reach this goal, we first state a necessary and sufficient condition for an edge in T' to belong to F_v. Then we show that the set of edges in T' satisfying this condition is connected.

To state the condition, we first observe that v induces a partition of ∂T given by $\partial T = D_1 \cup D_2 \cup D_3$. If e_1, e_2 and e_3 are as above, then e_1 induces the partition $\partial T = D_1 \cup (D_2 \cup D_3)$, e_2 induces the partition $\partial T = D_2 \cup (D_1 \cup D_3)$ and e_3 induces the partition $\partial T = (D_1 \cup D_2) \cup D_3$. We claim that an edge e' in T' belongs to F_v if and only if neither of the sets e'^+ and e'^- is entirely contained in any of the D_is. Note that the hypothesis $(**)$ above implies that it is not possible to have $e'^+ = D_i$ or $e'^- = D_i$ for any i. We go on to prove the claim.

One direction is straightforward: in fact if e'^+ (or e'^-) is contained in one of the D_is, then the partition induced by e' and the one induced by e_j would be nested for each $j = 1, 2, 3$. Let us prove now that if for each $i = 1, 2, 3$ we have $e'^+ \not\subseteq D_i$ and $e'^- \not\subseteq D_i$, then there exists an i such that the partition induced by e' and the one induced by e_i are not nested, i.e. there exists an i such that all the sets $e'^+ \cap D_i$, $e'^- \cap D_i$, $e'^+ \cap {D_i}^C$, $e'^- \cap {D_i}^C$ are non-empty. To prove this, note first that there exists an i such that both of the sets $e'^+ \cap D_i$ and $e'^- \cap D_i$ are non-empty; in fact, if this were not true, there would exist an i such that either $e'^+ = D_i$ or $e'^- = D_i$, contradicting hypothesis $(**)$. Now, since e'^+ is not entirely contained in D_i, the set $e'^+ \cap {D_i}^C$ is also non-empty, and since e'^- is not entirely contained in D_i, the set $e'^- \cap {D_i}^C$ is non-empty, which proves the claim.

To prove that the set of edges in T' satisfying this condition is connected, we use a similar argument as in the proof of Lemma 9.4.5: we show that if two edges a and b in T' belong to F_v then the geodesic segment in T' joining a and b is contained in F_v. Suppose $a^+ \supset b^+$. Then for any edge c in the geodesic segment joining a and b, we have $a^+ \supset c^+ \supset b^+$ and $a^- \subset c^- \subset b^-$. Since b^+ is not contained in any of the D_is, neither is c^+, and since a^- is not contained in any of the D_is, neither is c^-, i.e. c belongs to F_v, which concludes the proof. □

Connectedness of the complex $C(T,T')$ is now an obvious consequence of Lemma 9.4.8, Lemma 9.4.9, and Lemma 9.4.5. This concludes the proof of Proposition 9.4.4.

Another consequence of Lemma 9.4.5 and Lemma 9.4.9 is the following:

Proposition 9.4.10 *The complex C is simply connected.*

Proof Any loop l in C would project, by compactness, to a finite subtree of T. Therefore, in order to prove that C is simply connected, it is sufficient to show that, for any finite subtree S of T, the preimage $\pi_T^{-1}(S)$ (which we denote by F_S), is simply connected.

Now note that if S is a finite subtree of T, then F_S is the union over all the edges e and all the vertices v in S of the preimages F_e and F_v, which are all simply connected by Lemma 9.4.5 and Lemma 9.4.9. Moreover, distinct preimages intersect, if at all, in a finite tree. Now, simple connectedness of F_S can be proven by using induction on the number of edges in S and Van Kampen's theorem. □

As a consequence of Proposition 9.4.10, the fundamental group of the quotient $\Delta(T,T')$ is the free group F_g.

Next we try to understand what the vertex links in $C(T,T')$ look like. First note that they are all subgraphs of the bipartite graph $K_{3,3}$, since the complex $C(T,T')$ is contained in the product $T \times T'$. We show next that all possible vertex links in $C(T,T')$ are the nine graphs described in Figure 9.3, which coincide with the nine graphs listed in Figure 9.1.

Consider a vertex (v,v') in $T \times T'$, and let us try to understand what its link in $C(T,T')$ is (this link is empty in the case where (v,v') is not in $C(T,T')$).

Denote by e_1, e_2 and e_3 (resp. e_1', e_2' and e_3') the three edges incident to v in T (resp. to v' in T'). Recall that v induces a partition $\partial T = D_1 \cup D_2 \cup D_3$, where the edge e_1 induces the partition $\partial T = D_1 \cup (D_2 \cup D_3)$; the edge e_2 induces the partition $\partial T = D_2 \cup (D_1 \cup D_3)$; the edge e_3 induces the partition $\partial T = (D_1 \cup D_2) \cup D_3$. The same holds for the vertex v' in T'. Now (v,v') is incident to nine squares in $T \times T'$, and understanding what the link of (v,v') looks like boils down to understanding which of the squares $e_i \times e_j$ belong to the core, i.e. for which i's and j's the partitions induced by e_i and e_j are non-nested. Namely, the link of the vertex (v,v') will consist of two sets of at most three vertices: a black set representing the edges e_1, e_2 and e_3 and a red set representing the edges e_1', e_2' and e_3', where the ith black vertex and the jth red vertex

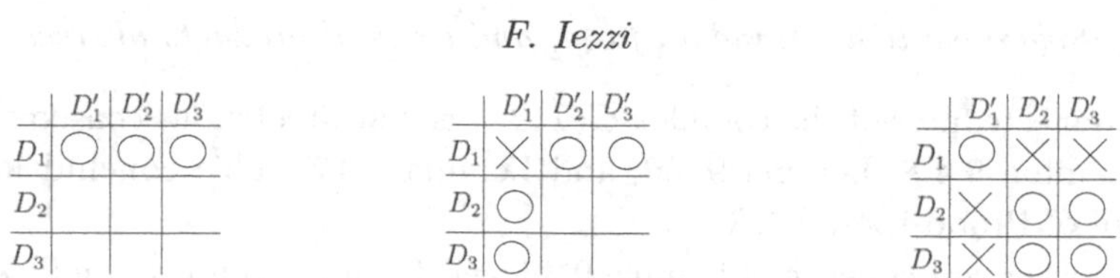

Figure 9.2 The patterns drawn above are "forbidden" as vertex tables. In fact, the pattern on the left hand side would imply that ∂T is empty; the pattern in the centre would imply $D_1 = D'_1$, consequently e_1 and e'_1 would induce the same partition, contradicting hypothesis $(**)$; the pattern on the right hand side would imply $D_1 = D'_2 \cup D'_3$, again contradicting hypothesis $(**)$.

are adjacent if and only if the partitions induced by the edges e_i and e'_j are non-nested.

We can deduce the link of (v, v') by drawing and analysing a simple 3×3 table associated to (v, v'), as shown in Figure 9.3. The table has a cross in the slot (i, j) if the set $D_i \cap D'_j$ is non-empty, and a circle in the slot (i, j) if the set $D_i \cap D'_j$ is empty. In the caption to Figure 9.3 we explain how to deduce from the position of crosses and circles whether the partitions induced by the edges e_i and e'_j are nested, for $i, j = 1, 2, 3$.

It is not difficult to systematically analyse all the possible vertex tables. These are 3 by 3 tables whose entries can be only crosses or circles. Moreover, they have to satisfy some additional conditions: first, since $\partial T = D_1 \cup D_2 \cup D_3 = D'_1 \cup D'_2 \cup D'_3$, there has to be at least a cross in each row and column of the table; second, by hypothesis $(**)$, the union of a row and a column must contain at least two crosses (see Figure 9.2 for an example of these "forbidden patterns"). Furthermore, permuting the order of rows or columns in the table or reflecting the table through the diagonal would not change the vertex link.

Figure 9.3 gives an exhaustive list of all such 3 by 3 tables up to permutation of rows or columns and reflection through the diagonal, and the vertex links associated to each table.

Summarising, given two trivalent trees T and T' endowed with actions by the group F_g, the core $C(T, T')$ satisfies the following properties:

(1) $C(T, T')$ is a connected V-H locally finite square complex;

(2) hyperplanes in $C(T, T')$ are finite trees;

(3) all possible vertex links in $C(T, T')$ are the nine graphs listed in Figure 9.3; in particular, no vertex link contains a triangle, which implies that $C(T, T')$ is locally CAT(0);

(4) $C(T, T')$ is simply connected and endowed with a free properly-discontinuous action of the free group F_g;

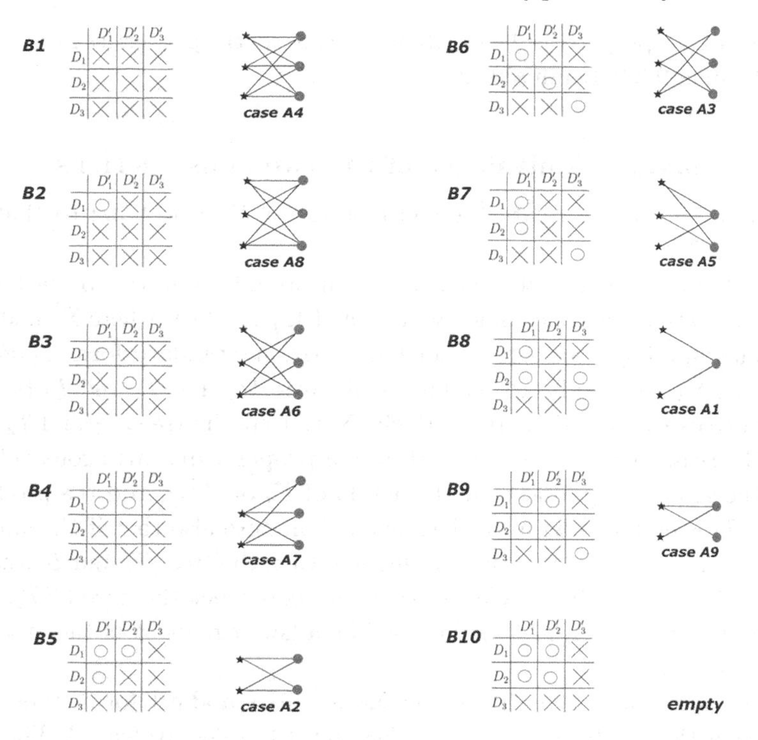

Figure 9.3 This figure describes all the possible vertex tables. As mentioned, for a vertex (v, v') in $T \times T'$ we draw a 3×3 table. The slot (i, j) contains a cross if the set $D_i \cap D'_j$ is non-empty and a circle otherwise. The partitions induced by the edges e_i and e'_j are non-nested if and only if the table corresponding to (v, v') satisfies the following four properties: the slot (i, j) contains a cross; the row i contains at least another cross; the column j contains at least another cross; the complement of the row i and the column j contains at least one cross. At the right hand side of each vertex table we draw the vertex link. We relate the graphs listed here to the ones listed in Figure 9.1.

(5) collapsing the vertical or horizontal lines to points yields projections of $C(T, T')$ to two trivalent trees.

On the other hand, the complex $\Delta(T, T')$ satisfies properties 2) and 3), and the following:

(1') Δ is a connected V-H finite square complex;

(4') the fundamental group of $\Delta(T, T')$ is the free group F_g;

(5') collapsing the vertical or horizontal lines to points yields projections of $\Delta(T, T')$ to two trivalent graphs.

Note that properties (1)–(5) above are exactly the properties mentioned in Lemma 9.3.2–Lemma 9.3.6.

9.4.2 Equivalence of the two constructions

In this section we relate the construction of Section 9.3.2 to that of Section 9.4.

Note first that the above construction provides a more abstract way to associate a square complex to a triple $(M_g, \Sigma_1, \Sigma_2)$, where Σ_1 and Σ_2 are maximal sphere systems not necessarily in standard form. Namely, let $\widetilde{\Sigma}_1$, $\widetilde{\Sigma}_2$ be the lifts to the universal cover $\widetilde{M}_g$. Let T_1 and T_2 be the dual trees to $\widetilde{\Sigma}_1$ and $\widetilde{\Sigma}_2$ respectively. Note that the trees T_1 and T_2 are both trivalent and endowed with a free properly-discontinuous action of the group F_g (induced by the action of F_g on $\widetilde{M}_g$), and the product $T_1 \times T_2$ is endowed with the diagonal action; note also that if Σ_1 and Σ_2 satisfy hypothesis (*) in the introduction, then the trees T_1 and T_2 satisfy hypothesis (**) above. We can therefore construct the core $C(T_1, T_2)$, and the quotient $\Delta(T_1, T_2)$. This will be a key ingredient in the proof of Theorem 9.6.1.

We show next that if Σ_1 and Σ_2 are in standard form, then the constructions of Section 9.3.2 and Section 9.4 are equivalent. A different proof of this result can be found in [Ho] (Proposition 2.1).

Proposition 9.4.11 *With the above notation, if Σ_1 and Σ_2 are in standard form, then the complex $\Delta(\widetilde{M}_g, \widetilde{\Sigma}_1, \widetilde{\Sigma}_2)$ defined in Section 9.3.2 is isomorphic to the core $C(T_1, T_2)$.*

Proof For the remainder of the proof we denote the square complex $\Delta(\widetilde{M}_g, \widetilde{\Sigma}_1, \widetilde{\Sigma}_2)$ by $\widetilde{\Delta}$. We first claim that $\widetilde{\Delta}$ can be identified with a subcomplex of the product $T_1 \times T_2$. In fact, collapsing the vertical (resp. horizontal) lines to points yields equivariant projections $p_1 : \widetilde{\Delta} \to T_1$ (resp. $p_2 : \widetilde{\Delta} \to T_2$): a square in $\widetilde{\Delta}$, representing an intersection between a sphere σ_1 in $\widetilde{\Sigma}_1$ and a sphere σ_2 in $\widetilde{\Sigma}_2$ gets projected through p_1 to the edge $e_1 \in T_1$ representing σ_1, and through p_2 to the edge $e_2 \in T_2$ representing σ_2. Moreover, since (by standard form) a sphere in $\widetilde{\Sigma}_1$ and a sphere in $\widetilde{\Sigma}_2$ can intersect at most once, then given edges $e_1 \in T_1$ and $e_2 \in T_2$ there is at most one square s in $\widetilde{\Delta}$ so that $p_1(s)$ is e_1 and $p_2(s)$ is e_2. Hence the claim holds.

To conclude the proof we show that each square in $\widetilde{\Delta}$ is contained in $C(T_1, T_2)$ and vice versa. Let σ_1 be a sphere in $\widetilde{\Sigma}_1$, let σ_2 be a sphere in $\widetilde{\Sigma}_2$, and let e_1 and e_2 be the edges of T_1 and T_2 representing σ_1 and

σ_2 respectively. Now the square $e_1 \times e_2$ is in $\widetilde{\Delta}$ if and only if $\sigma_1 \cap \sigma_2$ is non-empty, if and only if (Lemma 9.2.2) the partitions induced by σ_1 and σ_2 on $End(\widetilde{M}_g)$ are non-nested, if and only if the partitions induced by e_1 and e_2 on the boundary $\partial T_1 = \partial T_2$ are non-nested, if and only if $e_1 \times e_2$ is in $C(T_1, T_2)$. □

As a consequence, the complex $\Delta(\widetilde{M}_g, \widetilde{\Sigma}_1, \widetilde{\Sigma}_1)$ satisfies properties (1)–(5) above. Hence Lemmas 9.3.2–9.3.6 hold.

Note now that the complex $\Delta(\widetilde{M}_g, \widetilde{\Sigma}_1, \widetilde{\Sigma}_2)$ is endowed with a free properly-discontinuous action of the free group F_g induced by the action of F_g on the manifold $\widetilde{M}_g$, and that if we see $\Delta(\widetilde{M}_g, \widetilde{\Sigma}_1, \widetilde{\Sigma}_1)$ as a subcomplex of $T_1 \times T_2$, then such an F_g-action coincides with the (restriction of the) diagonal action of F_g on $T_1 \times T_2$. Therefore an immediate consequence of Proposition 9.4.11 is the following:

Corollary 9.4.12 *The square complex $\Delta(M_g, \Sigma_1, \Sigma_2)$ is isomorphic to the square complex $\Delta(T_1, T_2)$.*

As a consequence of Corollary 9.4.12, the complex $\Delta(M_g, \Sigma_1, \Sigma_2)$ satisfies properties (2)–(3), (1')–(5') above.

Proposition 9.4.11 shows that the square complex dual to two maximal sphere systems in standard form can always be realised as the core of two trivalent trees. In the next section we will prove a sort of converse, i.e. the core of two trivalent trees both endowed with actions of the free group F_g can always be realised as the square complex dual to two maximal sphere systems embedded in the manifold $\widetilde{M}_g$.

9.5 The inverse construction

In Section 9.3.2, given a manifold M_g with two embedded sphere systems in standard form, we have decomposed the manifold as a union of simple pieces, we have described a dual square complex and proved it satisfies properties (2), (3), (1')–(5') of Section 9.4. In this section we are going to describe an inverse procedure. Namely, starting with a square complex that satisfies the above properties, we will associate to each cell of the complex a "piece", and we will glue these pieces together according to some "gluing rules". We will then show that this provides a piece decomposition for a manifold M_g and two embedded sphere systems in standard form.

Let Δ be a square complex satisfying properties (2), (3), (1')–(5') of Section 9.4. We associate a piece to each cell of Δ in the following way.

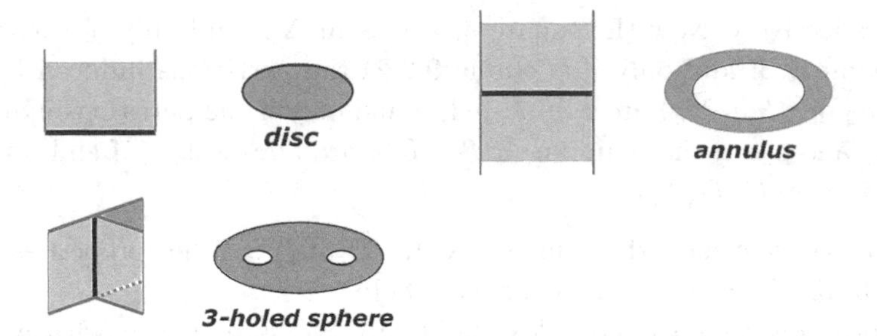

Figure 9.4 How to associate a 2-piece $p(e)$ to an edge e. The edge we consider is the thicker edge in the picture. The associated 2-piece is a disc if the edge bounds one square, an annulus if the edge bounds two squares, and a pair of pants if the edge bounds three squares.

Given a square s in Δ we associate to s a circle, $c(s)$. We call these circles *1-pieces.*

As for edges, we will refer to horizontal edges as *black edges* and to vertical edges as *red edges.* We associate to each edge e in Δ a planar surface $p(e)$ having as many boundary components as the number of squares adjacent to e (which is at most 3 by property (5')). See Figure 9.4. We colour these surfaces with black or red, according to the colour of the edge they are associated to, and we call these surfaces *2-pieces.*

We associate to each vertex v a closed orientable surface $S(v)$ and a handlebody $P(v)$ according to the link of v in Δ. The surface $S(v)$ is the union of the 2-pieces and the 1-pieces associated to the edges and squares incident to v; the vertex link determines how these pieces are glued together, as described in Figure 9.5; gluing maps are meant to be self-homeomorphisms of the circle. Note that it is always possible to choose the gluing maps in such a way that $S(v)$ is orientable, and that (the homeomorphism class of) $S(v)$ is uniquely determined. Then $P(v)$ is the (orientable) handlebody having $S(v)$ as its boundary and (in case $S(v)$ has positive genus) such that 1-pieces on $S(v)$ are not all trivial in the fundamental group of $P(v)$.

We call the handlebody $P(v)$ a *3-piece* and the surface $S(v)$ the boundary pattern of $P(v)$. Note that two 2-pieces of the same colour are never adjacent on the boundary of a 3-piece.

With a little abuse we will use the word 2-piece (resp 3-piece) to denote both the open and the closed surface (resp. handlebody).

We now glue these pieces together to form a topological space, which we denote by M_Δ; we will show (Lemma 9.5.2) that M_Δ is a 3-manifold.

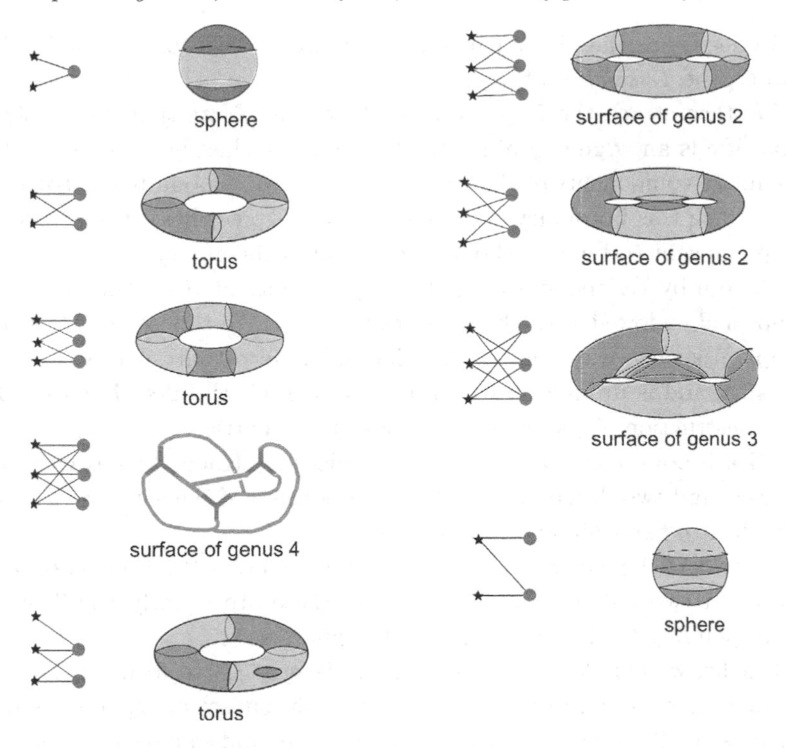

Figure 9.5 How to deduce the boundary pattern $S(v)$ (on the right) given the link of the vertex v (on the left): if G is the link of a vertex v in Δ, each vertex of valence k in G corresponds to a planar surface having k boundary components (where k is 1, 2 or 3); two surfaces are glued along a circle if and only if the two corresponding vertices are joined by an edge in the link G; we choose the gluing maps in such a way that the resulting boundary pattern is orientable. The 3-piece $P(v)$ is the exterior of the surfaces we draw (if $S(v)$ has positive genus we require that 1-pieces on the boundary are not all trivial in $\pi_1(P(v))$). (The darker surfaces correspond to black 2-pieces; the paler surfaces correspond to red 2-pieces.)

The space M_Δ will be constructed inductively, first taking the union of the 1-pieces, then attaching the 2-pieces and eventually the 3-pieces. The idea is that we attach an n-piece to an $(n-1)$-piece if the corresponding cells of the square complex are incident to each other. The procedure is described below.

Let N_s be an indexing set for the squares of the complex Δ, let N_e be an indexing set for the edges of Δ, and let N_v be an indexing set for the vertices of Δ.

Define C_1 as the disjoint union of the circles $c(s_i)$ for all i in N_s. We call C_1 the *1-skeleton* of M_Δ.

We then attach the 2-pieces to the 1-skeleton. Note that, by construction, if e is an edge contained in the square s, then exactly one of the boundary components of the 2-piece $p(e)$ will correspond to the square s. We attach this boundary component to the 1-piece $c(s)$; the attaching map is meant to be a self-homeomorphism of the circle.

Denote by C_2 the space $C_1 \bigsqcup_{i \in N_e} p(e_i)$ quotiented by the attaching maps and endowed with the quotient topology. We choose the attaching maps in such a way that C_2 is orientable. Note that such a choice is possible and is unique up to isotopy. We call C_2 the *2-skeleton* of M_Δ. By construction, C_2 satisfies the following properties:

– Each boundary component of a 2-piece is attached to exactly one 1-piece, and two different boundary components of the same 2-piece are attached to two different 1-pieces.

– For each 1-piece $c = c(s)$, exactly four different 2-pieces, two black ones and two red ones, are glued to c; these are exactly the 2-pieces corresponding to the four edges of the square s.

Finally, we glue the 3-pieces to the 2-skeleton C_2 to form M_Δ.

Namely, given a vertex v in Δ, consider the subset of C_2 composed of 1-pieces and 2-pieces which correspond to edges and squares in the star of v. Note that this is a connected closed orientable surface, and coincides with the boundary pattern of the 3-piece $P(v)$. Therefore we can glue the 3-piece $P(v)$ to its boundary pattern in C_2: recall that, if $P(v)$ is not a ball, we glue it in such a way that the 1-pieces in the boundary pattern are not all trivial in $\pi_1(P(v))$. Here the gluing map is defined only up to Dehn twists around curves on the boundary pattern that are homotopic to 1-pieces (i.e. essential curves in the 2-pieces). For the moment we choose the gluing maps and carry on with the construction. We will observe below (Remark 9.5.4) that a different choice for the gluing maps would in the end give us a homeomorphic 3-manifold, and therefore our choice is not relevant.

Denote the union of C_2 and the 3-pieces, quotiented by the attaching maps, by M_Δ. Denote the union of 1-pieces and black 2-pieces by Q_B, and the union of 1-pieces and red 2-pieces by Q_R.

Note that each 2-piece lies on the boundary of exactly two 3-pieces and each 1-piece lies on the boundary of exactly four 3-pieces.

In the same way, if $\widetilde{\Delta}$ denotes the universal cover of such a square complex Δ (equivalently, $\widetilde{\Delta}$ satisfies properties (1)–(5) in Section 9.4), we can construct a topological space $M_{\widetilde{\Delta}}$. Denote by $\widetilde{Q}_B$ the union of

1-pieces and black 2-pieces in $M_{\widetilde{\Delta}}$, and by $\widetilde{Q}_R$ the union of 1-pieces and red 2-pieces in $M_{\widetilde{\Delta}}$.

Note that the action of the group F_g on $\widetilde{\Delta}$ induces a free properly-discontinuous cocompact action of F_g on $M_{\widetilde{\Delta}}$. We deduce that $M_{\widetilde{\Delta}}$ is a covering space for M_Δ, and F_g is the deck transformation group for the covering map; $\widetilde{Q}_R$ and $\widetilde{Q}_B$ are the full lifts of Q_R and Q_B. We will prove (Lemma 9.5.7), that $M_{\widetilde{\Delta}}$ is in fact the universal cover of M_Δ.

The rest of this section is aimed at proving the following:

Theorem 9.5.1 *The space M_Δ is the connected sum of g copies of $S^2 \times S^1$. Q_R and Q_B are two embedded maximal sphere systems in standard form with respect to each other.*

The proof of Theorem 9.5.1 consists of several steps. First we prove:

Lemma 9.5.2 *M_Δ is a closed topological 3-manifold.*

Proof We claim that each point q in M_Δ has a neighbourhood homeomorphic to $\mathbb{R}^3$. To prove it, we analyse separate cases.

The claim is clearly true if the point q belongs to the interior of a 3-piece.

Suppose the point q belongs to the interior of a 2-piece p. Now, p lies on the boundary of exactly two 3-pieces, P_1 and P_2, which are glued along p. Therefore there exists a neighbourhood U_1 of q in P_1 homeomorphic to ${\mathbb{R}^3}_-$, and a neighbourhood U_2 of q in P_2 homeomorphic to ${\mathbb{R}^3}_+$, so that U_1 and U_2 are glued together along their common boundary; their union provides a neighbourhood of q in M_Δ homeomorphic to $\mathbb{R}^3$.

Finally, suppose that q is contained in a 1-piece c. The piece c lies on the boundary of exactly four 2-pieces and exactly four 3-pieces. Again, by choosing suitable neighbourhoods of q in the four 3-pieces it belongs to and gluing them together, we can find a neighbourhood of q in M_Δ homeomorphic to $\mathbb{R}^3$.

This proves that M_Δ is a 3-manifold without boundary. Now M_Δ is compact because it is a finite union of 3-pieces. □

In the same way, we can show that $M_{\widetilde{\Delta}}$ is a topological 3-manifold, though not necessarily compact. As a next step we prove the following:

Lemma 9.5.3 *Each connected component of Q_B or Q_R is an embedded sphere in M_Δ.*

Proof By construction, each 1-piece is an embedded circle in M_Δ and each 2-piece is an embedded surface; two different 2-pieces are either disjoint or they are glued together along a 1-piece, and each 1-piece bounds exactly two red 2-pieces and two black 2-pieces. Consequently, Q_B and Q_R are embedded surfaces in M_Δ without boundary, possibly disconnected.

Note now that two 2-pieces of the same colour are glued together along a 1-piece if and only if the edges they correspond to are the two horizontal (or vertical) edges of the same square, i.e. there is a bijective correspondence between the hyperplanes perpendicular to black (resp. red) edges and the connected components of Q_B (resp. Q_R).

There is indeed a systematic way to recover components of Q_B or Q_R from the hyperplane it corresponds to. Namely, if we consider a hyperplane H as a graph embedded in $\mathbb{R}^3$, then the corresponding surface will be the boundary of a tubular neighbourhood of H, which is a sphere since by property 2) H is a finite tree. □

Note that Lemma 9.5.3 holds also for $\widetilde{Q}_R$ and $\widetilde{Q}_B$ in $M_{\widetilde{\Delta}}$.

Remark 9.5.4 As promised, we observe now that, if we had chosen different gluing maps for the 3-pieces, the manifold we obtained would be homeomorphic to M_Δ, i.e. M_Δ is well-defined. As mentioned above, two attaching maps for a 3-piece differ by a composition of Dehn twists around curves on the 2-pieces. Now suppose M_Δ and M'_Δ have been constructed from the same square complex Δ following the above procedure, and suppose they differ by only one gluing map : $P(v) \to C_2$, where $P(v)$ is a 3-piece and C_2 is the 2-skeleton; call the two different gluing maps $f : P(v) \to C_2$ and $f' : P(v) \to C_2$. Suppose also that f and f' differ by a single Dehn twist around a curve γ on $\partial P(v)$, where γ is entirely contained in a 2-piece. We can assume $f(\gamma)$ and $f'(\gamma)$ are identified with the same curve in the 2-skeleton C_2, call this curve δ. Then one can check that the manifold M'_Δ can be obtained from M_Δ by performing a Dehn surgery of kind $(1, 1)$ on δ. By Lemma 9.5.3, the curve δ lies on an embedded sphere in M_Δ, hence it bounds an embedded disc, which implies that M'_Δ is homeomorphic to M_Δ. We can now conclude by induction. The same is true for $M_{\widetilde{\Delta}}$.

As a next step towards the proof of Theorem 9.5.1, we claim that the fundamental group of M_Δ is the free group F_g of rank g.

We have already observed that the manifold $M_{\widetilde{\Delta}}$ is a covering space for M_Δ and the deck transformation group is the free group F_g. We will prove that $M_{\widetilde{\Delta}}$ is simply connected, which implies that $M_{\widetilde{\Delta}}$ is the universal cover of M_Δ, hence the claim holds.

The proof of simple connectedness of $M_{\widetilde{\Delta}}$ relies on some preliminary lemmas.

Lemma 9.5.5 *For each 3-piece P in M_Δ or $M_{\widetilde{\Delta}}$, its fundamental group $\pi_1(P)$ is supported on the 1-piece components of its boundary pattern; i.e. there exists a basis for $\pi_1(P)$ so that each generator γ is homotopic to a 1-piece in the boundary pattern.*

Proof The proof is a case-by-case inspection. □

Lemma 9.5.6 *For each 3-piece P in M_Δ (resp. $M_{\widetilde{\Delta}}$), the inclusion $P \to M_\Delta$ (resp. $P \to M_{\widetilde{\Delta}}$) induces the trivial map on the level of fundamental groups.*

Proof By Lemma 9.5.5, each element of $\pi_1(P)$ can be represented as a product of loops each of which is homotopic to a 1-piece. By Lemma 9.5.3, each 1-piece lies on a sphere in M_Δ or $M_{\widetilde{\Delta}}$, therefore is trivial in $\pi_1(M_\Delta)$ (or $\pi_1(M_{\widetilde{\Delta}})$). □

The construction we describe next will be a very useful tool in the proof of simple-connectedness of $M_{\widetilde{\Delta}}$ as well as for the proof of Lemma 9.5.8 below.

To fix terminology, we use the term *binary subdivision* of a V-H square complex to describe the union of the 1-skeleton and the hyperplanes. We will show that a graph G isomorphic to the binary subdivision of $\widetilde{\Delta}$ can be embedded into $M_{\widetilde{\Delta}}$, and that each loop in the graph G is trivial in the fundamental group of $M_{\widetilde{\Delta}}$.

To build this graph G in $M_{\widetilde{\Delta}}$, we first build a subgraph G' that is isomorphic to the 1-skeleton of $\widetilde{\Delta}$. Then we add the other vertices and edges.

We construct G' as a dual to the piece decomposition of $M_{\widetilde{\Delta}}$: take a point q_P in the interior of each 3-piece of $M_{\widetilde{\Delta}}$; join two points q_{P_1} and q_{P_2} by an edge α_p if the 3-pieces P_1, P_2 the points belong to both contain the 2-piece p on their boundary. We require α_p to be an embedded arc and to intersect only one 2-piece (the 2-piece p) in exactly one point. We colour the edges in G' with black or red, each edge inheriting the colour of the 2-piece it intersects. We use the term *4-circuits* to denote loops in G' consisting of the concatenation of four edges. Note that G' is, by construction, isomorphic to the 1-skeleton of $\widetilde{\Delta}$ as a coloured graph; note also that 4-circuits coincide with loops of minimal length, and that each loop in G' can be written as a concatenation of 4-circuits (since both the horizontal and vertical projection of $\widetilde{\Delta}$ are trees).

Now, to build G from G' we take a new vertex for each 4-circuit in G' and join it to the midpoints of the four edges composing the circuit. Namely, consider a 4-circuit in the graph G. This circuit corresponds to four 2-pieces p_1, p_2, p_3 and p_4 in $M_{\widetilde{\Delta}}$, all intersecting in a 1-piece c. Denote by α_i the edge in G' dual to the 2-piece p_i. Take a point q in the circle c, and for each $i = 1...4$ take an arc β_i entirely contained in the 2-piece p_i and joining the point q to the arc α_i (see Figure 9.6). Call the edges in $G \setminus G'$ *newedges* and colour each newedge with black or red, according to the colour of the 2-piece it lies on. We will use the word *bisectors* to denote the union of newedges belonging to the same component $\widetilde{Q}_R$ or $\widetilde{Q}_B$. Note that each bisector is an embedded tree in a component of $\widetilde{Q}_R$ or $\widetilde{Q}_B$ and that bisectors correspond exactly to hyperplanes in $\widetilde{\Delta}$.

By construction, each 4-circuit of G (i.e. each loop in G consisting of four edges) is entirely contained in a single 3-piece of $M_{\widetilde{\Delta}}$, and therefore is trivial in $\pi_1(M_{\widetilde{\Delta}})$, by Lemma 9.5.6. Moreover, since G is isomorphic to the binary subdivision of $\widetilde{\Delta}$ whose vertical and horizontal projections are trees, each loop in G can be written as a concatenation of 4-circuits. Consequently, each loop in G (and in particular each loop in G') is nullhomotopic in $M_{\widetilde{\Delta}}$. We are now ready to prove the following:

Lemma 9.5.7 *$M_{\widetilde{\Delta}}$ is simply connected.*

Proof We have constructed above a graph G embedded in $M_{\widetilde{\Delta}}$, and we have shown that each circuit in G is trivial in $\pi_1(M_{\widetilde{\Delta}})$. We show now that each loop l in $M_{\widetilde{\Delta}}$ is homotopic to a loop in the graph G, which implies the lemma.

Let l be a loop in $M_{\widetilde{\Delta}}$. Up to homotopy, we can suppose that l does not intersect any 1-piece, and intersects every 2-piece transversely. We may as well suppose that l intersects a 2-piece p, if at all, in the point $p \cap G'$ (which is a vertex of G by construction). Now, by Lemma 9.5.6, if P is any 3-piece in $M_{\widetilde{\Delta}}$, then $l \cap P$ can be homotoped into $G \cap P$. □

As a consequence of Lemma 9.5.7, the fundamental group of M_Δ is the free group F_g. Now, M_Δ is a compact topological 3-manifold without boundary. Using 5.2 and 5.3 in [He] together with Perelman's solution to the Poincaré conjecture, we deduce that M_Δ is the connected sum of g 2-sphere bundles over S^1. Since M_Δ is orientable, it is a connected sum of g copies of $S^2 \times S^1$.

To complete the proof of Theorem 9.5.1, we only need to show that

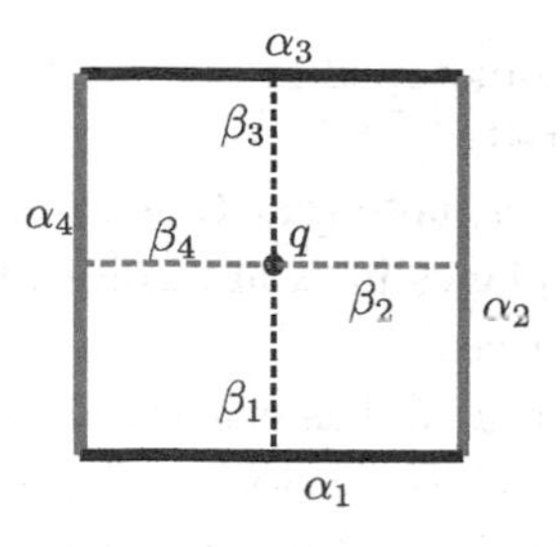

Figure 9.6 A 4-circuit in the graph G' and the newedges. The dotted lines represent the newedges.

Q_R and Q_B are maximal sphere systems in standard form. To reach this goal, we first prove a preliminary Lemma:

Lemma 9.5.8 *All the complementary components of $\widetilde{Q}_B$ and all the complementary components of $\widetilde{Q}_R$ in $M_{\widetilde{\Delta}}$ are 3-holed 3-spheres.*

Proof For the sake of simplicity, we will work in this proof with the closure of the components of $M_{\widetilde{\Delta}} \smallsetminus \widetilde{Q}_B$ and $M_{\widetilde{\Delta}} \smallsetminus \widetilde{Q}_R$. Let C be (the closure of) a component of $M_{\widetilde{\Delta}} \smallsetminus \widetilde{Q}_B$; note that C is a 3-manifold with boundary and its boundary consists of a certain number of spheres. Using the construction of the graph G described above, we show that C is simply connected, compact, and has exactly three boundary components. We deduce, using Perelman's solution to the Poincaré conjecture, that C is a 3-holed 3-sphere. By symmetry, the same can be proven for any complementary component of $\widetilde{Q}_R$.

Note first that C is a union of 3-pieces, 2-pieces and 1-pieces of $M_{\widetilde{\Delta}}$. Here each 1-piece is contained in a boundary component of C, hence Lemma 9.5.5 implies that for each 3-piece P in C the inclusion $P \to C$ induces the trivial map on the level of fundamental groups. Note also that if $G \subset M_{\widetilde{\Delta}}$ is the graph we constructed above, then the intersection between a component of $\widetilde{Q}_B$ and the graph G is exactly a black bisector in the graph G. Hence the intersections between G and the components of $M_{\widetilde{\Delta}} \smallsetminus \widetilde{Q}_B$ are the complementary components of black bisectors in G. Denote by G_C the graph $C \cap G$.

Now each loop in the graph G_C is nullhomotopic in C (since each 4-circuit is entirely contained in a 3-piece, and each loop in G_C can be written as a concatenation of 4-circuits). Moreover, using the same argument as in the proof of Lemma 9.5.7, we can show that any loop l in C is homotopic to a loop in the graph G_C (namely, we can suppose that l meets each 2-piece in a vertex of G_C, then we observe that for any

3-piece P, $l \cap P$ can be homotoped into $G_C \cap P$). This suffices to deduce that C is simply connected.

To see that C is compact, note that G is locally finite and bisectors in G are finite (since hyperplanes in $\widetilde{\Delta}$ are finite); therefore C is the union of a finite number of 3-pieces.

To conclude, we show that C has exactly three boundary components. To this aim, note that black bisectors are perpendicular to black edges, hence, property (5) in Section 9.4 implies that collapsing each red edge and each black bisector in the graph G to a point yields a trivalent tree, i.e. each complementary component of black bisectors in G is bounded by exactly three bisectors. As a consequence, C is bounded by exactly three spheres. □

We can now prove the following.

Lemma 9.5.9 *Each component of Q_R and Q_B is an essential sphere in M_Δ. Further, Q_R and Q_B are maximal sphere systems in M_Δ, in standard form with respect to each other.*

Proof As usual, denote the full lifts of Q_R and Q_B to the universal cover $M_{\widetilde{\Delta}}$ by $\widetilde{Q}_R$ and $\widetilde{Q}_B$. By Lemma 9.5.8 each component of $M_{\widetilde{\Delta}} \setminus \widetilde{Q}_B$ and each component of $M_{\widetilde{\Delta}} \setminus \widetilde{Q}_R$ is a 3-holed 3-sphere; consequently each component of $M_\Delta \setminus Q_B$ and each component of $M_\Delta \setminus Q_R$ is a 3-holed 3-sphere, which implies that each sphere in Q_R or Q_B is essential and that Q_R and Q_B are maximal sphere systems.

Now recall that Q_R and Q_B being in standard form means that Q_R and Q_B are in minimal form (i.e. each sphere in $\widetilde{Q}_B$ intersects each sphere in $\widetilde{Q}_R$ minimally) and that all complementary components of $Q_R \cup Q_B$ in M_Δ are handlebodies.

The latter condition is satisfied by construction.

To see that Q_R and Q_B are in minimal form, note that components of $\widetilde{Q}_B$ and $\widetilde{Q}_R$ correspond to hyperplanes in $\widetilde{\Delta}$. Now $\widetilde{\Delta}$ is simply connected and locally CAT(0); therefore it is CAT(0), by a generalisation of the Cartan–Hadamard Theorem ([BH] p. 193). Since two hyperplanes in a CAT(0) cube complex intersect at most once, a component of $\widetilde{Q}_R$ and a component of $\widetilde{Q}_B$ intersect at most once in $M_{\widetilde{\Delta}}$. Moreover, by construction, no 3-piece in $M_{\widetilde{\Delta}}$ is bounded by two disks. These two facts imply that each sphere in $\widetilde{Q}_R$ intersects each sphere in $\widetilde{Q}_B$ minimally. □

This concludes the proof of Theorem 9.5.1.

Note that the sphere systems Q_R and Q_B also satisfy property (*) in the introduction. In fact, since each edge in Δ bounds a square (by property (3) in Section 9.4), each sphere in Q_R must intersect Q_B and vice versa.

Remark 9.5.10 Note also that the construction we described above is in some sense "inverse"to the one we described in Section 9.3.2, as we explain below.

If we apply the construction described in Section 9.3.2 to the manifold M_Δ (resp. $M_{\widetilde{\Delta}}$), then we obtain the complex Δ (resp. $\widetilde{\Delta}$), i.e. Δ (resp. $\widetilde{\Delta}$) is the dual square complex associated to (M_Δ, Q_B, Q_R) (resp. to ($M_{\widetilde{\Delta}}$, $\widetilde{Q}_B$, $\widetilde{Q}_R$)). As a consequence, the horizontal and vertical projections of $\widetilde{\Delta}$ are the dual trees to $\widetilde{Q}_B$ and $\widetilde{Q}_R$ in $M_{\widetilde{\Delta}}$. This fact will be a key ingredient in the proof of Theorem 9.6.1.

On the other hand, if (Σ_1, Σ_2) is a pair of embedded maximal sphere systems in M_g in standard form, and $\Delta(M_g, \Sigma_1, \Sigma_2)$ is the dual square complex, then applying the above construction to $\Delta(M_g, \Sigma_1, \Sigma_2)$ yields a 3-manifold M_Δ, with a pair (Q_B, Q_R) of maximal sphere systems in standard form. It can be easily seen that there exists also a homeomorphism $F : M_g \to M_\Delta$ mapping the pair (Σ_1, Σ_2) to the pair (Q_B, Q_R). In fact there is a bijective correspondence between the pieces of $(M_g, \Sigma_1, \Sigma_2)$ and the pieces of (M_Δ, Q_B, Q_R) (since both sets correspond to the cells of Δ), and this correspondence respects the gluing relation (i.e. an $n-1$-piece of $(M_g, \Sigma_1, \Sigma_2)$ lies on the boundary of an n-piece of $(M_g, \Sigma_1, \Sigma_2)$ if and only if the same is true for the corresponding pieces of (M_Δ, Q_B, Q_R)). The map F then maps each piece of $(M_g, \Sigma_1, \Sigma_2)$ to the corresponding piece of (M_Δ, Q_B, Q_R) homeomorphically.

9.6 Consequences and applications

Before proving the main theorems we give a summary of what we have done so far. In Section 9.3.2 we have shown a constructive way to associate a square complex to a triple $(M_g, \Sigma_1, \Sigma_2)$, where Σ_1 and Σ_2 are maximal sphere systems in standard form. In Section 9.4 we have described a more abstract and general construction. This construction allows us to build a square complex starting with a triple $(M_g, \Sigma_1, \Sigma_2)$, where Σ_1 and Σ_2 are maximal sphere systems, not necessarily in standard form. The square complex will be (a quotient of) the core of $T_1 \times T_2$ (where, for $i = 1, 2$, T_i is the dual tree to $\widetilde{M}_g$ and $\widetilde{\Sigma}_i$, endowed with the group action induced by the F_g-action on $\widetilde{M}_g$). Proposition 9.4.11 then shows that the two constructions are equivalent in the case that Σ_1 and Σ_2 are in standard form. In Section 9.5 we have then described a sort of inverse, wherein, starting with a square complex, we build the manifold M_g with two maximal sphere systems in standard form. The constructions described in Section 9.4 and in Section 9.5 are the main ingredients of the following:

Theorem 9.6.1 *Let M_g be the connected sum of g copies of $S^2 \times S^1$ and let Σ_1, Σ_2 be two embedded maximal sphere systems which satisfy hypothesis (*) in the introduction. Then there exist maximal sphere systems (Σ_1', Σ_2') such that Σ_i' is homotopic to Σ_i for $i = 1, 2$, and Σ_1', Σ_2' are in standard form.*

Before proving Theorem 9.6.1 we clarify some terminology: given two infinite trivalent trees T and T' endowed with an identification of their boundaries, we say that T and T' *coincide* if there exists a simplicial isomorphism $\varphi : T \to T'$ such that for each edge e in T its image $\varphi(e)$ induces the same partition as e on the boundary. We are now ready to prove Theorem 9.6.1.

Proof Let $\widetilde{\Sigma}_1$ and $\widetilde{\Sigma}_2$ be the full lifts of Σ_1 and Σ_2 to $\widetilde{M}_g$ and let T_1 and T_2 be the dual trees to $\widetilde{\Sigma}_1$ and $\widetilde{\Sigma}_2$ respectively. Note that T_1 and T_2 are trivalent trees and they are both endowed with a geometric action of the group F_g, induced by the action of F_g on $\widetilde{M}_g$. As mentioned in Section 9.4, the group action induces an identification of the boundaries of T_1 and T_2. Note that, since Σ_1 and Σ_2 satisfy hypothesis (*), the trees T_1 and T_2 satisfy hypothesis (**) in Section 9.4. Let $C(T_1, T_2)$ be the core of $T_1 \times T_2$. By applying the construction of Section 9.5 to $C(T_1, T_2)$, we obtain a triple $(M_C, \widetilde{Q}_B, \widetilde{Q}_R)$ endowed with a geometric action of the group F_g, inherited from the group action on $C(T_1, T_2)$. Here M_C is homeomorphic to $\widetilde{M}_g$ and $\widetilde{Q}_B$, $\widetilde{Q}_R$ are two embedded maximal sphere systems in standard form with respect to each other. By construction, $C(T_1, T_2)$ is the dual square complex to the triple $(M_C, \widetilde{Q}_B, \widetilde{Q}_R)$. Hence the vertical and horizontal projection of $C(T_1, T_2)$, i.e. T_1 and T_2, coincide with the dual trees to $\widetilde{Q}_B$ and $\widetilde{Q}_R$ respectively.

The space of ends of M_C can be identified with the space of ends of $\widetilde{M}_g$, as both can be identified with the boundaries of T_1 and T_2. Moreover, since the tree dual to $\widetilde{M}_g$ and $\widetilde{\Sigma}_1$ coincides to the tree dual to M_C and $\widetilde{Q}_B$ (they both coincide with the tree T_1), for each sphere σ in $\widetilde{\Sigma}_1$ there is a sphere in $\widetilde{Q}_B$ inducing the same partition as σ on the space of ends, and for each sphere s in $\widetilde{Q}_B$ there is a sphere in $\widetilde{\Sigma}_1$ inducing the same partition as s. The same holds for $\widetilde{\Sigma}_2$ and $\widetilde{Q}_R$.

We can find an F_g-equivariant homeomorphism $H : M_C \to \widetilde{M}_g$ which is consistent with the identification on the space of ends (inherited from identifying both spaces of ends with the boundaries of T_1 and T_2). Denote $H(\widetilde{Q}_B)$ by $\widetilde{\Sigma}'_1$ and $H(\widetilde{Q}_R)$ by $\widetilde{\Sigma}'_2$.

The systems $\widetilde{\Sigma}'_1$ and $\widetilde{\Sigma}'_2$ are maximal and are in standard form with respect to each other, since they are homeomorphic images of two maximal sphere systems in standard form. Moreover, for each sphere in $\widetilde{\Sigma}_1$ (resp. $\widetilde{\Sigma}_2$) there is a sphere in $\widetilde{\Sigma}'_1$ (resp. $\widetilde{\Sigma}'_2$) inducing the same partition on the space of ends and vice versa. Hence, by Lemma 9.2.1, for $i = 1,2$ the sphere system $\widetilde{\Sigma}_i$ is homotopic in $\widetilde{M}_g$ to the sphere system $\widetilde{\Sigma}'_i$.

Let Σ'_1 and Σ'_2 in M_g be the projections of $\widetilde{\Sigma}'_1$ and $\widetilde{\Sigma}'_2$ through the covering map. These are two embedded maximal sphere systems in M_g in standard form with respect to each other, and moreover for $i = 1,2$, the sphere system Σ'_i is homotopic in M_g to the sphere system Σ_i. □

To summarise, in the proof of Theorem 9.6.1 we have shown a constructive way to find a standard form for two maximal sphere systems in M_g. Note that Theorem 9.6.1 can also be proven using the existence of Hatcher's normal form (Proposition 1.1 in [Ha]).

Another consequence of the construction of Section 9.5 is a kind of uniqueness result for standard form. More precisely:

Theorem 9.6.2 *Let (Σ_1, Σ_2), (Σ'_1, Σ'_2) be two pairs of embedded maximal sphere systems in M_g. Suppose that both pairs of sphere systems are in standard form and satisfy hypothesis $(\star)$ in the introduction. Suppose also that Σ_i is homotopic to Σ'_i for $i = 1,2$. Then there exists a homeomorphism $F : M_g \to M_g$ such that $F(\Sigma_i) = \Sigma'_i$ for $i = 1,2$. The homeomorphism F induces an inner automorphism of the fundamental group of M_g.*

The proof of Theorem 9.6.2 is based on some preliminary lemmas.

First we show that, as a consequence of the constructions described in Section 9.3.2 and in Section 9.5, a pair of sphere systems in standard form is somehow determined by its dual square complex. Namely:

Lemma 9.6.3 *Let (Σ_1, Σ_2), (Σ'_1, Σ'_2) be two pairs of embedded maximal sphere systems in M_g, both in standard form with respect to each other. Suppose the dual square complexes $\Delta(M_g, \Sigma_1, \Sigma_2)$ and $\Delta(M_g, \Sigma'_1, \Sigma'_2)$ are isomorphic as coloured square complexes. Then there exists a homeomorphism $H : M_g \to M_g$ so that $H(\Sigma_i)$ is Σ'_i for $i = 1,2$.*

Proof If we denote by (M_Δ, Q_R, Q_B) (resp. (M'_Δ, Q'_R, Q'_B)) the triple obtained by applying the construction of Section 9.5 to $\Delta(M_g, \Sigma_1, \Sigma_2)$

(resp. to $\Delta(M_g, \Sigma'_1, \Sigma'_2)$), then, by construction and by Remark 9.5.4, there is a homeomorphism $M_\Delta \to M'_\Delta$ mapping the pair (Q_B, Q_R) to the pair (Q'_B, Q'_R). Hence, Lemma 9.6.3 immediately follows from Remark 9.5.10. □

The next result is well known. Since, however, we did not find a reference, we give a proof below.

Lemma 9.6.4 *For $g \geq 3$, let $F : M_g \to M_g$ be a self-homeomorphism of M_g. Let Σ be a maximal sphere system in M_g. Suppose that the image $F(\sigma)$ of each sphere σ in Σ is homotopic to σ. Then the induced homomorphism $F_* : \pi_1(M_g) \to \pi_1(M_g)$ is an inner automorphism of the free group F_g.*

Proof As usual, denote the universal cover of M_g by $\widetilde{M}_g$, and denote the full lift of Σ by $\widetilde{\Sigma}$. The manifold $\widetilde{M}_g$ is endowed with an action of the free group F_g and the quotient of $\widetilde{M}_g$ by this action is the manifold M_g. In order to prove Lemma 9.6.4 we will show that a lift $\widetilde{F}$ of the homeomorphism F is equivariant under this group action.

To this aim, first note that any homeomorphism $H : \widetilde{M}_g \to \widetilde{M}_g$ is determined, up to homotopy, by its behaviour on the spheres in $\widetilde{\Sigma}$ (since each component of $\widetilde{M}_g \setminus \widetilde{\Sigma}$ is a 3-holed 3-sphere), and induces a homeomorphism $H_E : End(\widetilde{M}_g) \to End(\widetilde{M}_g)$.

Note also that the F_g-action on $\widetilde{M}_g$ induces an action of F_g on the space of ends; on the other hand, this action on the space of ends determines the action on $\widetilde{M}_g$ up to homotopy (in fact, since each component of $\widetilde{M}_g \setminus \widetilde{\Sigma}$ is a 3-holed 3-sphere, the action of F_g on $\widetilde{M}_g$ is determined by the action of F_g on $\widetilde{\Sigma}$, which is determined up to homotopy by the action of F_g on the space of ends of $\widetilde{M}_g$). Consequently, a homeomorphism $H : \widetilde{M}_g \to \widetilde{M}_g$ is F_g-equivariant (up to homotopy) if and only if the induced map $H_E : End(\widetilde{M}_g) \to End(\widetilde{M}_g)$ is equivariant under the induced F_g-action on $End(\widetilde{M}_g)$.

Now let $\widetilde{\sigma}$ be a sphere in $\widetilde{\Sigma}$. Since F fixes the homotopy class of each sphere in Σ, we can choose $\widetilde{F} : \widetilde{M}_g \to \widetilde{M}_g$ in such a way that the sphere $\widetilde{F}(\widetilde{\sigma})$ is homotopic to the sphere $\widetilde{\sigma}$ in $\widetilde{M}_g$.

In addition, the image $\widetilde{F}(\widetilde{\sigma})$ determines the image $\widetilde{F}(\widetilde{\tau})$ for each τ in $\widetilde{\Sigma}$ (here we are using that, since $g \geq 3$, a triple of spheres in Σ bounds at most one component of $M_g \setminus \Sigma$). This implies that $\widetilde{F}$ fixes the homotopy class of each sphere in $\widetilde{\Sigma}$, and therefore for each $\widetilde{\tau}$ in $\widetilde{\Sigma}$, the sphere $\widetilde{F}(\widetilde{\tau})$ induces the same partition on the space of ends of $\widetilde{M}_g$ as the sphere $\widetilde{\tau}$.

As a consequence, the homeomorphism $\widetilde{F}_E : End(\widetilde{M}_g) \to End(\widetilde{M}_g)$

induced by $\widetilde{F}$ is equivariant under the F_g-action on $End(\widetilde{M_g})$, which implies that $\widetilde{F}$ is equivariant up to homotopy under the F_g-action on $\widetilde{M_g}$. □

Remark 9.6.5 Lemma 9.6.4 holds also in the case where g is 2, under the additional hypothesis that F fixes the components of $M_g \setminus \Sigma$ up to homotopy; in particular it holds when F is orientation preserving.

We are now ready to prove Theorem 9.6.2.

Proof of Theorem 9.6.2 Let T_1, T_2, T_1', T_2' be the dual trees to $\widetilde{\Sigma}_1$, $\widetilde{\Sigma}_2$, $\widetilde{\Sigma}_1'$ and $\widetilde{\Sigma}_2'$ respectively. Since, for $i = 1,2$, the system Σ_i is homotopic to the system Σ_i', the core $C(T_1, T_2)$ is isomorphic as a V-H square complex to the core $C(T_1', T_2')$, and the quotients $\Delta(T_1, T_2)$ and $\Delta(T_1', T_2')$ are also isomorphic. Thus, by Proposition 9.4.11, the square complex dual to $(M_g, \Sigma_1, \Sigma_2)$ is isomorphic as a coloured square complex to the square complex dual to $(M_g, \Sigma_1', \Sigma_2')$. Now, by Lemma 9.6.3, there exists a homeomorphism $F : M_g \to M_g$ such that the image $F(\Sigma_i)$ is Σ_i', for $i = 1,2$. Further, since F is induced by the isomorphism of dual square complexes, F fixes the homotopy class of each sphere in Σ_1 and Σ_2, and of each component of $M_g \setminus \Sigma_1$. Hence, by Lemma 9.6.4 (and by Remark 9.6.5 if $g = 2$), the map F induces an inner automorphism of the fundamental group F_g. □

We conclude this section with the following:

Remark 9.6.6 A theorem by Laudenbach ([L2] page 80) states that if $Mod(M_g)$ denotes the group of (isotopy classes of) self-homeomorphisms of the manifold M_g and $H : Mod(M_g) \to Out(F_g)$ is the homomorphism sending a map to its action on $\pi_1(M_g)$, then the kernel of this map is the subgroup of $Mod(M_g)$ generated by a finite number of sphere twists (namely twists around spheres in a maximal sphere system). In light of this result, we can restate Theorem 9.6.2 in the following way:

Two standard forms for a pair (Σ_1, Σ_2) of maximal sphere systems, can be obtained one from the other by a combination of self-homeomorphisms of M_g homotopic to the identity, and twists around spheres in Σ_1.

References

[BBC] Jason Behrstock, Mladen Bestvina, and Matt Clay. Growth of intersection numbers for free group automorphisms. *J. Topol.*, 3(2):280–310, 2010.

[BH] Martin R. Bridson and André Haefliger. *Metric spaces of non-positive curvature*, volume 319 of *Grundlehren der Mathematischen Wissenschaften*. Springer-Verlag, Berlin, 1999.

[G] Vincent Guirardel. Cœur et nombre d'intersection pour les actions de groupes sur les arbres. *Ann. Sci. École Norm. Sup. (4)*, 38(6):847–888, 2005.

[Ha] Allen Hatcher. Homological stability for automorphism groups of free groups. *Comment. Math. Helv.*, 70(1):39–62, 1995.

[HV] Allen Hatcher and Karen Vogtmann. The complex of free factors of a free group. *Quart. J. Math. Oxford Ser. (2)*, 49(196):459–468, 1998.

[He] John Hempel. *3-manifolds*. AMS Chelsea Publishing, Providence, RI, 2004. Reprint of the 1976 original.

[HOP] Sebastian Hensel, Damian Osajda, and Piotr Przytycki. Realisation and dismantlability. *Geom. Topol.*, 18(4):2079–2126, 2014.

[HH] Arnaud Hilion and Camille Horbez. The hyperbolicity of the sphere complex via surgery paths. *Journal fr die reine und angewandte Mathematik*, 2012.

[Ho] Camille Horbez. Sphere paths in outer space. *Algebr. Geom. Topol.*, 12(4):2493–2517, 2012.

[I] F Iezzi. Sphere systems in 3-manifolds and arc graphs. *PhD thesis, University of Warwick*, 2016.

[L1] François Laudenbach. Sur les 2-sphères d'une variété de dimension 3. *Ann. of Math. (2)*, 97:57–81, 1973.

[L2] François Laudenbach. *Topologie de la dimension trois: homotopie et isotopie*. Société Mathématique de France, Paris, 1974. Astérisque, No. 12.

[P] Georg Peschke. The theory of ends. *Nieuw Arch. Wisk. (4)*, 8(1):1–12, 1990.

[Sa] Michah Sageev. CAT(0) cube complexes and groups. Unpublished notes:
`http //www.math.utah.edu/pcmi12/lecture_notes/sageev.pdf`.

[St] John R. Stallings. Whitehead graphs on handlebodies. In *Geometric group theory down under (Canberra, 1996)*, pages 317–330. de Gruyter, Berlin, 1999.

[W] J.H.C. Whitehead. On Certain Sets of Elements in a Free Group. *Proc. London Math. Soc.*, S2-41(1):48.

10

Uniform quasiconvexity of the disc graphs in the curve graphs

Kate M. Vokes

Mathematics Institute
University of Warwick
Coventry CV4 7AL
United Kingdom

Abstract

We give a proof that there exists a universal constant K such that the disc graph associated to a surface S forming a boundary component of a compact, orientable 3-manifold M is K-quasiconvex in the curve graph of S. Our proof does not require the use of train tracks.

10.1 Introduction

Given a closed, connected, orientable surface S, the associated *curve graph*, $\mathcal{C}(S)$, has as its vertex set the set of isotopy classes of essential simple closed curves in S, with an edge between two distinct vertices if the corresponding isotopy classes have representatives that are disjoint. The definition is due to Harvey [Har]. The curve graph has been a significant tool in the study of mapping class groups, Teichmüller spaces and hyperbolic 3-manifolds. If S has genus at least 2, then $\mathcal{C}(S)$ is connected, and Masur and Minsky showed in [MM1] that it has infinite diameter and is hyperbolic in the sense of Gromov. More recently, it was shown independently by Aougab [A1], Bowditch [B], Clay, Rafi and Schleimer [CRS] and Hensel, Przytycki and Webb [He] that the constant of hyperbolicity can be chosen to be independent of the surface S. If S has genus 1, then no two distinct isotopy classes of curves can be realised disjointly. In this case, it is usual to modify the definition so that curves are adjacent in $\mathcal{C}(S)$ if they intersect exactly once. The resulting graph is the *Farey graph*, which is connected. As for the higher genus cases, it is hyperbolic with infinite diameter.

When S is a boundary component of a compact, orientable 3-manifold

M, we can consider the subset of the vertex set of $\mathcal{C}(S)$ that consists of those curves which bound embedded discs in M. Equivalently, by Dehn's lemma, these are the essential simple closed curves in S that are homotopically trivial in M. The *disc graph*, $\mathcal{D}(M,S)$, is the full subgraph spanned by these vertices. This has applications to the study of handlebody groups and Heegaard splittings. In a further paper of Masur and Minsky [MM2], it was proved that $\mathcal{D}(M,S)$ is K-quasiconvex in $\mathcal{C}(S)$ (see Section 2 for a definition), for some K depending only on the genus of S. The proof relies on a study of nested train track sequences. More specifically, to any pair of vertices of $\mathcal{D}(M,S)$, Masur and Minsky associate a sequence of curves in $\mathcal{D}(M,S)$, and a nested train track sequence whose vertex cycles are close in $\mathcal{C}(S)$ to the curves of this sequence. They prove that the sets of vertex cycles of nested train track sequences are quasiconvex in $\mathcal{C}(S)$, and the result follows.

This result was improved by Aougab, who showed in [A2] that the constants of quasiconvexity for nested train track sequences can be taken to be quadratic in the complexity of the surface, obtaining as a corollary that there exists a function $K(g) = O(g^2)$ such that $\mathcal{D}(M,S)$ is $K(g)$-quasiconvex in $\mathcal{C}(S)$, where g is the genus of S. That this bound can be taken to be uniform in the genus of S follows from work of Hamenstädt [Ham]. In Section 3 of [Ham], it is shown that the sets of vertex cycles of train track splitting sequences give unparametrised quasi-geodesics in $\mathcal{C}(S)$ with constants independent of the surface S. Along with the uniform hyperbolicity of the curve graphs, this implies that such subsets are uniformly quasiconvex in $\mathcal{C}(S)$. In this note, we give a direct proof of the uniform quasiconvexity of $\mathcal{D}(M,S)$ in $\mathcal{C}(S)$ without using train tracks.

Theorem 10.1.1 *There exists K such that for any compact, orientable 3-manifold M and boundary component S of M, the disc graph $\mathcal{D}(M,S)$ is K-quasiconvex in $\mathcal{C}(S)$.*

For the main case, where the genus of S is at least 2, this uses an observation that the disc surgeries of [MM2] give a path of "bicorn curves", as described by Przytycki and Sisto in [PS]. These were introduced by analogy with the "unicorn arcs" of [He] in order to give a short surgery proof of the uniform hyperbolicity of the curve graphs. The lower genus case is straight-forward, and is discussed in Section 10.3.

Acknowledgements.
I am grateful to my supervisor, Brian Bowditch, for many helpful discussions and suggestions and for reading drafts of this article. I would also like to thank Francesca Iezzi and Saul Schleimer for interesting conversations and the referee for helpful comments. This work was supported by an Engineering and Physical Sciences Research Council Doctoral Award.

10.2 Preliminaries

A simple closed curve in a surface S is said to be *essential* if it does not bound a disc in S. We will assume from now on that all curves are essential simple closed curves. Two curves α and β, intersecting transversely, are in *minimal position* if the number of their intersection points is minimal over all pairs α', β' isotopic to α, β respectively. This is equivalent to the condition that α and β do not form a bigon, that is, an embedded disc in S whose boundary is a union of one arc of each of α and β. Abusing notation, we shall also denote the isotopy class of a curve α by α. The intersection number, $i(\alpha, \beta)$, is the number of intersections between representatives of the isotopy classes of α and β that are in minimal position.

If Y is a subset of a geodesic metric space X, we denote the closed K-neighbourhood of Y in X by $N_X(Y, K)$. We say that Y is *K-quasiconvex* in X if, for any two points y and y' in Y, any geodesic in X joining y and y' is contained within $N_X(Y, K)$. The metric on the curve graph $\mathcal{C}(S)$ is given by setting each edge to have length 1, and we denote the distance between vertices α and β by $d_S(\alpha, \beta)$. This makes $\mathcal{C}(S)$ into a geodesic metric space.

Given two curves α and β in minimal position, a *bicorn curve* between α and β is a curve constructed from one arc a of α and one arc b of β such that a and b meet precisely at their endpoints (see Figure 10.1 and [PS]). Since α and β are in minimal position, all bicorn curves between them will be essential, as otherwise α and β would form a bigon.

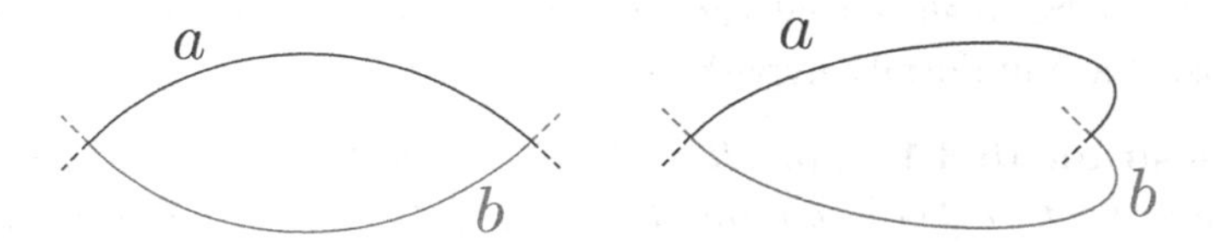

Figure 10.1 Examples of bicorn curves.

10.3 Exceptional cases

Let M be a compact, orientable 3-manifold. If a boundary component S has genus at most one then the associated disc graph $\mathcal{D}(M,S)$ is very simple. Firstly, since there are no essential simple closed curves on the sphere, the curve graph of the sphere is empty, so we can ignore any sphere boundary components. We shall see that for a torus boundary component S, $\mathcal{D}(M,S)$ contains at most one vertex.

Suppose S is a torus boundary component of the 3-manifold M. Suppose an essential curve δ in S bounds an embedded disc D in M. Take a closed regular neighbourhood N of $S \cup D$ in M. This is homeomorphic to a solid torus with an open ball removed. Suppose some other curve δ' in S bounds an embedded disc D' in M. We can assume that D' intersects the sphere boundary component of N transversely in simple closed curves. Repeatedly performing surgeries along innermost discs to reduce the number of such curves eventually gives a disc with boundary δ' which is completely contained in N. Therefore, an essential curve in S bounds an embedded disc in M if and only if it bounds an embedded disc in N. Up to isotopy, there is no curve in S other than δ that bounds an embedded disc in N, since such a curve must be trivial in the first homology group of N. We hence find that for any torus boundary component S, $\mathcal{D}(M,S)$ is at most a single point. In this case, $\mathcal{D}(M,S)$ is 0-quasiconvex, or convex, in the curve graph of S (which is the Farey graph, as described in the introduction).

10.4 Proof of the main result

Now let S be a boundary component of genus at least 2 of a compact, orientable 3-manifold M, and $\mathcal{D}(M,S)$ the associated disc graph.

The following criterion for hyperbolicity appears in several places in slightly different forms, for example as Theorem 3.15 of [MS]. For our purposes, the important result is the final clause on Hausdorff distances, which appears in the statement of Proposition 3.1 of [B]. This criterion is used in [PS] to show that the curve graphs of closed surfaces of genus at least 2 are uniformly hyperbolic.

Proposition 10.4.1 *For all $h \geq 0$, there exist k and R such that the following holds. Let G be a connected graph with vertex set $V(G)$. Suppose that for every $x, y \in V(G)$ there is a connected subgraph $\mathcal{L}(x,y) \subseteq G$, containing x and y, with the following properties:*

1 for any $x, y \in V(G)$ with $d_G(x,y) \le 1$, the diameter of $\mathcal{L}(x,y)$ in G is at most h,
2 for all $x, y, z \in V(G)$, $\mathcal{L}(x,y) \subseteq N_G(\mathcal{L}(x,z) \cup \mathcal{L}(y,z), h)$.

Then G is k-hyperbolic. Furthermore, for all $x, y \in V(G)$, the Hausdorff distance between $\mathcal{L}(x,y)$ and any geodesic from x to y is at most R.

We first note that a slightly more general statement is true.

Claim 10.4.2 *The requirement that each $\mathcal{L}(x,y)$ be connected can be replaced by the weaker condition that there exist h' such that, for all $x, y \in V(G)$, $N_G(\mathcal{L}(x,y), h')$ is connected.*

To see this, suppose subgraphs $\mathcal{L}(x,y)$ satisfy all the hypotheses of Proposition 10.4.1, except that the connectedness assumption is replaced as described in the claim. Define $\mathcal{L}'(x,y) = N_G(\mathcal{L}(x,y), h')$. This is a connected subgraph of G containing x and y. For any $x, y \in V(G)$ with $d_G(x,y) \le 1$, the diameter of $\mathcal{L}'(x,y)$ in G is at most $h + 2h'$, and for any $x, y, z \in V(G)$, $\mathcal{L}'(x,y) \subseteq N_G(\mathcal{L}'(x,z) \cup \mathcal{L}'(y,z), h)$. Hence, the conclusion of Proposition 10.4.1 holds, except with constants now depending on h and h', proving Claim 10.4.2.

Given two curves α and β in S, we shall define $\Theta(\alpha,\beta)$ to be the set containing the isotopy classes of α, β and all bicorn curves between α and β.

Przytycki and Sisto define in [PS] an "augmented curve graph", $\mathcal{C}_{aug}(S)$, in which two curves are adjacent if they intersect at most twice. Such curves cannot fill S (which has genus at least 2) so are at distance at most 2 in $\mathcal{C}(S)$. Given two curves α and β in minimal position, $\eta(\alpha,\beta)$ is defined in [PS] to be the full subgraph of $\mathcal{C}_{aug}(S)$ spanned by $\Theta(\alpha,\beta)$. This is shown to be connected for all α and β. It is further verified that the hypotheses of Proposition 10.4.1 are satisfied when G is $\mathcal{C}_{aug}(S)$, $\mathcal{L}(\alpha,\beta)$ is $\eta(\alpha,\beta)$ for each α, β, and h is 1, independently of the surface S.

Since $\eta(\alpha,\beta)$ is connected in $\mathcal{C}_{aug}(S)$, for any $\gamma, \gamma' \in \Theta(\alpha,\beta)$, there is a sequence $\gamma = \gamma_0, \gamma_1, \ldots, \gamma_n = \gamma'$ of curves in $\Theta(\alpha,\beta)$, where $d_S(\gamma_{i-1},\gamma_i) \le 2$ for each $1 \le i \le n$. Hence $N_{\mathcal{C}(S)}(\Theta(\alpha,\beta),1)$ is a connected subgraph of $\mathcal{C}(S)$. Moreover, if $d_S(\alpha,\beta) \le 1$, then α and β are disjoint, so $\Theta(\alpha,\beta)$ contains no other curves and its diameter in $\mathcal{C}(S)$ is at most 1. Finally, since $\eta(\alpha,\beta) \subset N_{\mathcal{C}_{aug}(S)}(\eta(\alpha,\delta) \cup \eta(\beta,\delta),1)$ for any curves α, β, δ, we have $\Theta(\alpha,\beta) \subset N_{\mathcal{C}(S)}(\Theta(\alpha,\delta) \cup \Theta(\beta,\delta),2)$.

Using Proposition 10.4.1 with the modification of Claim 10.4.2, this proves the following lemma.

Lemma 10.4.3 *There exists R such that, for any closed, orientable surface S of genus at least 2, and any curves α, β in S, the Hausdorff distance in $\mathcal{C}(S)$ between $\Theta(\alpha,\beta)$ and any geodesic in $\mathcal{C}(S)$ joining α and β is at most R.*

We now show that, moreover, any geodesic between α and β in $\mathcal{C}(S)$ lies in a uniform neighbourhood of any path within $\Theta(\alpha,\beta)$ connecting α and β.

Lemma 10.4.4 *Let α, β be two curves in S, $P(\alpha,\beta)$ a path from α to β in $\mathcal{C}(S)$ with all vertices in $\Theta(\alpha,\beta)$, and g a geodesic in $\mathcal{C}(S)$ joining α and β. Then g is contained in the $(2R+2)$-neighbourhood of $P(\alpha,\beta)$.*

Proof This uses a well-known connectedness argument. From Lemma 10.4.3, $P(\alpha,\beta)$ is contained in $N_{\mathcal{C}(S)}(g,R)$. Take any vertex γ in g. Let g_0 be the subpath of g from α to γ and let g_1 be the subpath from γ to β. Then the three sets $N_{\mathcal{C}(S)}(g_0,R+1)$, $N_{\mathcal{C}(S)}(g_1,R+1)$ and $P(\alpha,\beta)$ intersect in at least one vertex, say δ. Let γ_0 in g_0 and γ_1 in g_1 be such that $d_S(\gamma_0,\delta) \leq R+1$ and $d_S(\gamma_1,\delta) \leq R+1$. Now $d_S(\gamma_0,\gamma_1) \leq 2R+2$ and γ is in the (geodesic) subpath of g from γ_0 to γ_1, so $d_S(\gamma,\gamma_i) \leq R+1$ for either $i=0$ or $i=1$. Hence, $d_S(\gamma,\delta) \leq 2R+2$. Since γ was an arbitrary vertex in g and δ is in $P(\alpha,\beta)$, we have $g \subset N_{\mathcal{C}(S)}(P(\alpha,\beta),2R+2)$. □

Given that α and β bound embedded discs in M, we now describe how to choose $P(\alpha,\beta)$ so that all curves in the path are also vertices of $\mathcal{D}(M,S)$, following Section 2 of [MM2].

Assume curves α and β are fixed in minimal position and choose a subarc $J \subset \alpha$. Masur and Minsky define several curve replacements, of which we shall need only the following. A *wave curve replacement* with respect to (α,β,J) is the replacement of α and J by α_1 and J_1 as follows (see Figure 10.2). Let w be a subarc of β with interior disjoint from α, and endpoints p, q in the interior of J. Suppose that w meets the same side of J at both p and q; then w is called a *wave*. Let J_1 be the (proper) subarc of J with endpoints p, q, and define α_1 to be the curve $w \cup J_1$. This is an essential curve since α and β are in minimal position, so, in particular, no subarc of J and subarc of β can form a bigon. Where $\operatorname{int}(J) \cap \beta = \varnothing$, we define a curve replacement with respect to (α,β,J) by $\alpha_1 = \beta$, $J_1 = \varnothing$.

Remark 10.4.5 In [MM2], it is arranged that α_1 and β must intersect transversely and be in minimal position by requiring an additional condition on the wave w and by slightly isotoping $w \cup J_1$ to be disjoint from

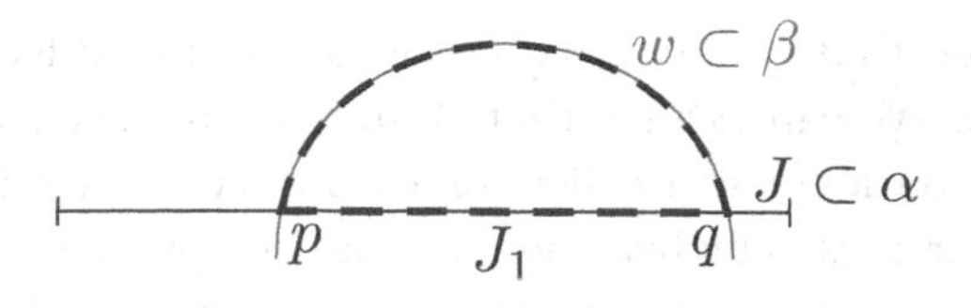

Figure 10.2 A wave curve replacement. The dashed curve is α_1.

w. However, this will not be necessary here, so we choose to simplify the exposition by removing this condition.

Notice that $i(\alpha, \alpha_1) = 0$ since α does not intersect $\mathrm{int}(w)$. Moreover, $\alpha_1 \cap \beta$ consists of the arc w and a set of points which are all contained in the interior of J_1, and $|\beta \cap \mathrm{int}(J_1)| < |\beta \cap \mathrm{int}(J)|$ whenever $|\beta \cap \mathrm{int}(J)|$ is non-zero.

We can iterate this process as follows. Although α_1 and β coincide in an arc, any intersections within the subarc J_1 are still transverse. Moreover, no subarc of J_1 can form a bigon with a subarc of β, since α and β are in minimal position. Hence, we may still define a wave curve replacement with respect to (α_1, β, J_1) as for (α, β, J) above and obtain an essential curve. A *nested curve replacement sequence* is a sequence $\{(\alpha_i, J_i)\}$ of curves $\alpha = \alpha_0, \alpha_1, \ldots, \alpha_n$ and subarcs $\alpha \supset J_0 \supset J_1 \supset \cdots \supset J_n$, such that J_0 contains all points of $\alpha \cap \beta$ in its interior, and such that α_{i+1} and J_{i+1} are obtained by a curve replacement with respect to (α_i, β, J_i). We will allow only wave curve replacements in the sequence and not the other curve replacements possible in [MM2]. We always have $i(\alpha_i, \alpha_{i+1}) = 0$, as for α and α_1. Observe that all curves α_i in this sequence are bicorn curves between α and β, since the nested arcs J_i ensure that they are formed from exactly one arc of α and one of β.

The following is a case of Proposition 2.1 of [MM2]. We include a proof for completeness, with the minor modification of the slightly different curve replacements.

Proposition 10.4.6 *Let S be a boundary component of a compact, orientable 3-manifold M, and let α and β be two curves in S in minimal position, each of which bounds an embedded disc in M. Let $J_0 \subset \alpha$ be a subarc containing all points of $\alpha \cap \beta$ in its interior. Then there exists a nested curve replacement sequence $\{(\alpha_i, J_i)\}$, with $\alpha_0 = \alpha$, such that:*

- *each α_i bounds an embedded disc in M,*
- *the sequence terminates with $\alpha_n = \beta$.*

Proof Suppose that α and β bound properly embedded discs A and B respectively. We can assume that A and B intersect transversely, so their intersection locus is a collection of properly embedded arcs and simple closed curves. Furthermore, we can remove any simple closed curve components by repeatedly performing surgeries along innermost discs, so that A and B intersect only in properly embedded arcs. We will perform surgeries on these discs to get a sequence of discs A_i with $\partial A_i = \alpha_i$. Throughout the surgeries, we will keep A and B fixed, and each A_i, except $A_0 = A$ and $A_n = B$, will be a union of exactly one subdisc of each of A and B.

Suppose the sequence is constructed up to $\alpha_i = \partial A_i$. If $\beta \cap \alpha_i$ is empty, then $\alpha_{i+1} = \beta = \partial B$ by definition, so the sequence is finished.

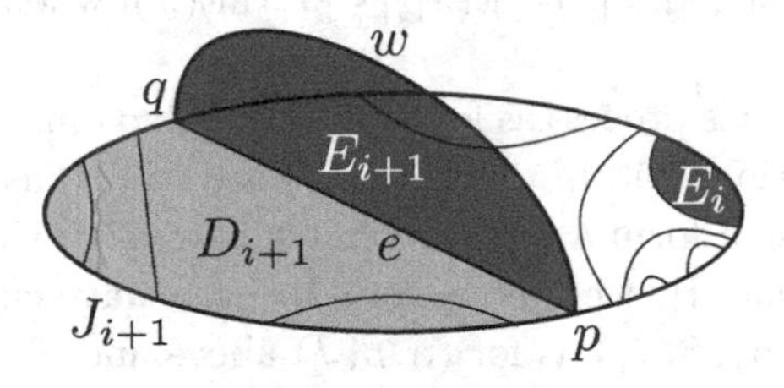

Figure 10.3 The disc surgeries of Proposition 10.4.6. The horizontal disc is A_i, shown with arcs of intersection with B.

Suppose β intersects α_i (as illustrated in the example of Figure 10.3). Let $A_i = D_i \cup E_i$, where D_i is a subdisc of A and E_i is a subdisc of B. If $i = 0$, then E_i is empty. If $i > 0$, let J_i be the arc of ∂D_i that is contained in ∂A_i. Any point of intersection of β and J_i is an endpoint of an arc of intersection of B and D_i. Let E_{i+1} be a disc in B such that the boundary of E_{i+1} is made up of an arc e in $\text{int}(D_i) \cap B$ and a subarc w of β, and such that the interior of E_{i+1} is disjoint from A_i (that is, E_{i+1} is an outermost component of $B \setminus (A_i \cap B)$). This in particular means that the interior of w is disjoint from α_i, that the endpoints p, q of w lie in the interior of J_i, and that w meets the same side of J_i at both of these endpoints, so w is a wave. Let J_{i+1} be the subarc of J_i with endpoints p, q. Let D_{i+1} be the disc in A_i bounded by $e \cup J_{i+1}$. This disc is contained in D_i and hence in A. The curve $w \cup J_{i+1}$, with interval J_{i+1}, is the wave curve replacement α_{i+1} obtained from (α_i, β, J_i), and it is also the boundary of the embedded disc $A_{i+1} = D_{i+1} \cup E_{i+1}$.

Since $|\beta \cap \text{int}(J_i)|$ decreases at each stage, this terminates with $|\beta \cap \text{int}(J_{n-1})| = 0$ and $\alpha_n = \beta$. $\square$

This sequence defines the vertices of a path $P(\alpha, \beta)$ in $\mathcal{C}(S)$, with these vertices contained in both $\mathcal{D}(M, S)$ and $\Theta(\alpha, \beta)$. By Lemma 10.4.4, there exists K, independent of S, α and β, such that any geodesic g joining α and β in $\mathcal{C}(S)$ is contained in the closed K-neighbourhood of $P(\alpha, \beta)$. Hence g is contained in the closed K-neighbourhood of $\mathcal{D}(M, S)$, completing the proof of Theorem 10.1.1.

References

[A1] Tarik Aougab, *Uniform hyperbolicity of the graphs of curves.* Geom. Topol. **17** (2013) 2855–2875.

[A2] Tarik Aougab, *Quadratic bounds on the quasiconvexity of nested train track sequences.* Topol. Proc. **44** (2014) 365–388.

[B] Brian H. Bowditch, *Uniform hyperbolicity of the curve graphs.* Pac. J. Math. **269** (2014) 269–280.

[CRS] Matt Clay, Kasra Rafi, Saul Schleimer, *Uniform hyperbolicity of the curve graph via surgery sequences.* Algebr. Geom. Topol. **14** (2014) 3325–3344.

[Ham] Ursula Hamenstädt, *Asymptotic dimension and the disk graph I.* To appear in *J. Topol.* Available at arXiv:1101.1843v4.

[Har] W.J. Harvey, *Boundary structure of the modular group.* In "Riemann surfaces and related topics: Proceedings of the 1978 Stony Brook conference", ed. I. Kra, B. Maskit, Ann. of Math. Stud. **97**, Princeton University Press (1981) 245–251.

[He] Sebastian Hensel, Piotr Przytycki, Richard C.H. Webb, *1-slim triangles and uniform hyperbolicity for arc graphs and curve graphs.* J. Eur. Math. Soc. **17** (2015) 755–762.

[MM1] Howard A. Masur, Yair N. Minsky, *Geometry of the complex of curves I: Hyperbolicity.* Invent. Math. **138** (1999) 103–149.

[MM2] Howard Masur, Yair Minsky, *Quasiconvexity in the curve complex,* in "In the tradition of Ahlfors and Bers, III", ed. W. Abikoff, A. Haas, Contemp. Math. **355**, Amer. Math. Soc. (2004) 309–320.

[MS] Howard Masur, Saul Schleimer, *The geometry of the disk complex,* J. Amer. Math. Soc. **26** (2013) 1–62.

[PS] Piotr Przytycki, Alessandro Sisto, *A note on acylindrical hyperbolicity of mapping class groups.* In "Hyperbolic geometry and geometric group theory", ed. K. Fujiwara, S. Kojima, K. Ohshika, *Adv. Stud. Pure Math.* **73**, Math. Soc. Japan, (2017), 255–264.